The Bold and the Brave:
A History of Women in Science and Engineering

D0253795

The Bold and the Brave:
A History of Women in Science and Engineering

MONIQUE FRIZE
WITH CONTRIBUTIONS FROM
PETER R. D. FRIZE
AND
NADINE FAULKNER

University of Ottawa Press

The University of Ottawa Press acknowledges with gratitude the support extended to its publishing list by Heritage Canada through its Book Publishing Industry Development Program, by the Canada Council for the Arts, by the Canadian Federation for the Humanities and Social Sciences through its Aid to Scholarly Publications Program, by the Social Sciences and Humanities Research Council, and by the University of Ottawa.

We also gratefully acknowledge the Faculty of Engineering and the office of the Vice President of research of the University of Ottawa as well as the Faculty of Engineering of the University of Carleton whose financial support has contributed to the publication of this book.

Library and Archives Canada Cataloguing in Publication

Frize, Monique, 1942–
The bold and the brave: a history of women in science
and engineering / Monique Frize; with contributions
from Peter R.D. Frize and Nadine Faulkner.

Includes bibliographical references and index.
ISBN 978-0-7766-0725-2

1. Women in science—History. 2. Women in engineering—
History. 3. Women—Education—History. 4. Women—
Intellectual life. 5. Women in science—Biography.
I. Frize, Peter R. D., 1940– II. Faulkner, Nadine, 1964–
III. Title.

Q130.F765 2009   305.43'5   C2009-906129-5

University of Ottawa Press
542 King Edward Avenue
Ottawa, Ontario K1N 6N5
www.uopress.uottawa.ca

uOttawa

To my son Patrick.

In memory of my mother, Paule,
and my sister, Danielle.

CONTENTS

PREFACE

> He decided to wait no longer before putting his project into effect, for he was afflicted by the thought of how much the world would suffer because of his tardiness. Many were the wrongs that had to be righted, grievances redressed, injustices made good, abuses removed, and duties discharged.
>
> —Cervantes, *Don Quixote*

This book represents, to a large extent, the legacy of the knowledge I acquired and shared while I was the first holder of the Northern Telecom (later Nortel) and Natural Sciences and Engineering Research Council (NSERC) chair for Women in Science and Engineering at the University of New Brunswick between December 1989 and June 1997. The chair was launched in May 1989 with a national mandate to help to increase the enrolment of women in engineering studies, and to encourage employers and the professions to adopt effective women-friendly policies in order to attract and retain increasing numbers of women in the professions and workplaces.

During the years when I was crisscrossing Canada to do this work, there were times when I felt like Don Quixote: there was so much to be done that I wondered if my mission was one of battling windmills. As I travelled from one province to another I met many women engineers and scientists who gave time to visiting schools and to mentoring a new generation of girls and young women. Many of these women also shared their expertise with me, examining the challenges and obstacles that limit the participation of women in some fields, especially physics, computer science, and electrical, mechanical, and computer engineering. The role of the chair was to be a catalyst, to network with leaders of dozens of initiatives in Canada, and elsewhere, to share best practices and ideas that were likely to be successful, and to discuss approaches that were seen to be less effective. There is no doubt that the initiatives launched across the country between 1990 and 1995 were very effective. The rate of women obtaining degrees in engineering in Canada in 1995 was 54 percent higher than it had been five years earlier (18.9 percent in 1995, compared to 14 percent in 1990).

Although recruitment strategies succeeded in helping to bring about a substantial increase in the numbers of women studying and working in science and engineering during the closing years of the 20th century, not only in Canada but in many other western countries, these gains were followed by a serious decline between 2003 and 2007. This decline occurred at a time when women were enrolling in larger numbers than men in post-secondary education, and when women made up more than half of the classes in medicine, law, and business. What was so peculiar about the physical sciences and engineering that they remain the only fields dominated by men in the early years of the 21st century? In order to understand why women are seriously underrepresented in a number of fields in science and engineering today, it is useful to understand the ways in which their participation in these fields has differed from men's participation throughout the ages. There have been brilliant women in science since the beginning of recorded history, but what were the educational opportunities for girls and women, compared to those for boys and men, in past centuries? How did society view women's intellectual capacities, and their ability to learn and do science and mathematics, throughout the ages? How did this affect their access to education, especially in the sciences and mathematics? Is the culture of science and engineering predominantly masculine? If so, how did this develop? Are there biases and prejudices involved in the ways in which success is rewarded, in what gets funded, and in what gets published? Finally, will a critical mass of women in science and engineering bring about some changes to the culture, to the topics of inquiry, and will women who join these fields be comfortable in recognizing and integrating so-called feminine values into the world of science and engineering? These are some of the questions explored in this book.

A major inspiration behind this book is the life and work of Christine de Pizan (1363–c. 1434), author of *Le Livre de la cité des dames* (1405, *The Book of the City of Ladies*), which depicts the lives and accomplishments of several women, including scientists and inventors, to demonstrate that women could be as learned as men. In addition to this remarkable achievement, she took up the battle against the misogyny portrayed in the poem the *Roman de la Rose*. Although the first part of this poem, written by Guillaume de Lorris (c. 1230), is an allegory of courtly love, the second part, added in 1275 by Jean de Meun, is a bawdy attack on women. This publication was the first French dialogue portraying the two views of women that lasted until the 17th century: women were seen as either romantic or vicious. Phyllis Stock (1978, pp. 41–42) describes Christine de Pizan's reaction when she read the *Roman de la Rose*: "It was not only that she opposed the misogyny of the second part of the work; she resented the general attitude of men toward women, reflected in the weakening of chivalry and the antifeminist preaching of the clergy." Her response was a poem, "L'Épistre au Dieu d'amours" (1399, "Epistle to the God of Love"), in which Cupid "presents to the other gods a women's petition asking for an end to the outrages they are forced to bear." The debate was taken up by many, on both sides of the issue, for quite some time, and helped to establish Christine de Pizan's reputation as a female intellectual who could assert herself effectively and defend her claims in the male-dominated literary realm. She continued to counter abusive literary treatments of women and was severely attacked for her audacity in criticizing a work of literature, the clergy, and men in general. However, she did receive some support from men such as Jean Gerson (1363–1429), a theologian, educator, and poet who was chancellor of the University of Paris (the Sorbonne), and regarded the *Roman de la Rose* as an immoral work. Throughout this book we meet several champions who, like Christine de Pizan, believed that women should have access to education and that they were not intellectually inferior to men. Alongside these supporters of women's education we also present the detractors, the men and women who were opposed to the education of women, believing that it would be wasted on them.

The purpose of this book, then, is to investigate how women have strived throughout history to gain access to education and careers in science and engineering. It introduces key concepts and debates in order to contextualize the obstacles that women have faced, and continue to face, in these fields.

The book is divided into four parts. Part I provides views of women's intellectual abilities, taken through the writings of thinkers, those who viewed women as inferior and those who did not, while paying attention to comparing persons who lived in similar eras. The discussion begins with ancient Greece, and continues with the Middle Ages, the Renaissance, and the Enlightenment. Part II discusses the education of women from the 17th century to the 19th century. The approach compares the views of men and women who lived in the same epochs, demonstrating that the belief in the superiority of men did not just flourish a long time ago but has existed and has been supported in every era, including the present day.

It will become clear from these two parts of the book that the voices raised against the education of women have been louder than the voices supporting women in every era until late in the 19th century. Those who believed that women were intellectually inferior explained the existence of a few brilliant women as mere exceptions to the rule. They appear not to have considered that these women had developed their abilities with the aid of tutors or relatives, and that many other women could have done the same if they too had been provided with the opportunities to develop their skills.

It will also become clear that women had more freedom to be educated, to participate in science and mathematics, and even to hold public roles in some periods than in others. Access to knowledge and involvement in science or mathematics has tended to move cyclically, with progress being followed by regress. There has not been a constant march toward equality, and there is clear evidence that women's full and equal participation in society is neither an automatic result of the simple passage of time nor in any way guaranteed. An acquaintance with women's history is the first step toward understanding and possibly developing ways to maintain gains and prevent future regresses. This has been a major impetus behind the writing of this book. If we all understand how things became the way they are today, we may be able to concentrate our efforts for as long as it takes to ensure that we see progress in our time and in the future. Another goal of this book is to render women more visible, since they have mostly been ignored by writers on the history of science. Even modern books on "eminent men of science" usually profile very few women, or none at all. Most troubling is that many people believe that biases and prejudices about women's abilities and skills are a thing of the past.

Part III begins with a description of the rising involvement of women in science and engineering in the 20th century, then presents the obstacles that have limited their progress, before reviewing successful recruitment and outreach programmes, strategies designed to achieve progress at all levels of education and career development, and the experiences of women in workplaces, where they have had to deal with the myth of "meritocracy" and the masculine culture of science. If we are to learn from history, we ought not to ignore the fact that, along with the great gains women have made in academia and science, there is still not a critical mass of women studying and working in these domains, and women are conspicuously absent from several new scientific fields, such as nanotechnology and robotics. The so-called pipeline model, which assumed that, if women filled the lower positions, they would eventually share the top positions, has been acknowledged as a failure (Schiebinger 1999, p. 14).

Finally, Part IV, written by Peter Frize, presents profiles of three bold and brave women whose contributions or potential did not get the recognition they deserved: Sophie Germain, Mileva Einstein, and Rosalind Franklin.

In looking at particular arguments both for and against women's equality, it becomes clear that rational argument alone has not been the liberator of women. Thus, while it is important to be able to recognize and uncover the weaknesses of bad arguments, it is equally important to realize that argument alone does not automatically result in change. What is more, disagreement about the similarities and differences between women and men, and what value we ought to place upon them, does not always mean that one side must be wrong and irrational and the other side right and rational. There is indeed much rational disagreement about the similarities and differences between men and women, and how these may affect their career choices and their lives. The aim here is not to decide these issues once and for all, but rather to provide readers with a frame of reference within which these arguments can be placed. This should also help to make sense of what is being said in current discourses on the subject.

—⚬✖✖⚬—

ACKNOWLEDGEMENTS

The author wishes to thank several people who have reviewed the manuscript at different stages and made helpful suggestions for improving it.

In particular, my thanks go to Peter Frize, my husband of forty years, for his support for my career, his patient reading of the manuscript (several times), and his writing of the last chapter. His suggestions have helped to make the book flow better, and to make it, we hope, more lively and more interesting.

I also wish to thank Nadine Faulkner, a graduate in philosophy who, from July 2000 to June 2001, did most of the research for the chapters on philosophers and for some of the material on the education of women. She also wrote the first drafts of the chapters discussing philosophers' views of women's minds, and provided a lot of comments on my first draft on the education of women. My thanks go to Debra Hauer, who helped to confirm references and formatted them. I truly appreciate her help in this important task.

My thanks also go to Ruby Heap, Professor of History at the University of Ottawa, who reviewed several of the sections on education in the earlier centuries and the story of Mileva Einstein. She provided excellent comments that helped to improve the manuscript. More recently, Dr. Eda Kranakis, a professor of history and technology, provided invaluable advice on how to make revisions to the manuscript in order to strengthen the historical sections, especially on the debate regarding the "scientific revolution."

I wish to thank all the women engineers and scientists who have believed in this project and encouraged me to complete it, with special recognition of the encouragement I have received from Claire Deschênes, Professor in Mechanical Engineering at the Université Laval, and from Misa Gratton, a retired mathematician and a former president of the Ottawa chapter of Women in Science and Engineering (WISE).

I must also thank my two universities, Carleton and Ottawa, for approving my two sabbatical leaves, a full year in 2003–04 and a half year in the winter term of 2008, which allowed me to complete the writing of the manuscript. My thanks also go to the Natural Sciences and Engineering Research Council, and to Nortel, for their financial support of the chair for Women in Science and Engineering (Ontario) between 1989 and 2002, and to Telesat Canada for the special grants provided in support of this project.

—❦—

PART I

---❊❊❊---

Views of Women's Intellectual Abilities

CHAPTER 1

⊶⊷

From Ancient Times
to Early Modern Europe

Through the ages, philosophers, almost all of whom were male, have affected society's thoughts as well as reflecting them. The discussion begins with one of the most influential philosophers in the western tradition, the ancient Greek philosopher Plato (423–347 BCE). Plato's influence through the ages, on many philosophical topics, has been enormous, but he is of interest here because of his radical views regarding women (see Bluestone 1994). In his well-known work *The Republic*, Plato argues that women, like men, can rule, and that those who demonstrates the required talents ought to be given access to education. Plato's views about women as expressed in this work do not reflect society's thoughts of the time, and they had no impact on how Greek society viewed women. The poor reception of his views, whether it was to argue against them or to ignore them, does, however, tell us something about the society we came from. In this chapter, we first look at the society in which Plato lived and study what he said about women. His view is then contrasted with Aristotle's and with the views of other thinkers throughout the ages.

THE STATUS OF WOMEN
IN ANCIENT GREECE

Plato lived in Athens, a city generally considered the birthplace of democracy. When seeking to judge whether a principle such as democracy applies to all people fairly, it is helpful to ask: for whom? The answer in ancient Greece was: not for everyone. The principle applied only to the Athenian male citizens, and excluded women, slaves, and those who were not native Athenians (Kitto 1951, pp. 124–25). Through their examination of the literature and art of the time, including philosophical essays, letters, political speeches, legal documents, plays, poems, and pottery, scholars have determined that Athenian women did not participate in politics, were generally excluded from socializing with men other than their husbands or close relatives, were not allowed to attend schools, could not own property, had limited freedom to move about alone, and generally lived under the care of a man, whether father, guardian, or husband (Kitto 1951, p. 218). However, this is not to say that women had no influence. Women contributed to Athenian society not only through their husbands but also because they were almost entirely responsible for the domestic aspect of life, which included not just bearing and raising children and managing households but also storing and preparing food, as well as crafts such as weaving and sewing. Similarly, slaves were an integral part of Athenian society (Davis 1978, pp. 99–103). Poorer women worked outside the home, selling their crafts or produce in the marketplace, or as midwives or wetnurses. There were also female physicians at this time.

Girls and women received some formal education at home, in subjects such as rhetoric (the art of persuasive speaking), from fathers, brothers, or husbands; and they learned domestic skills such as weaving, music, and dancing, all in preparation for their allotted social role. Boys went to public schools and studied grammar, rhetoric, arithmetic, music, geometry, and astronomy. The exceptions to the rule were prostitutes or women companions (*hetaerae*). Although legally no freer than other Athenian women, they associated more freely with men and had more freedom of movement. In order to be more interesting companions to men, they attended schools and were taught grammar, rhetoric, and dialectic (a form of argumentation). One well-known example of an extraordinary woman, thought by some scholars to have been a *hetaera*, is Aspasia of Miletus (c. 470–c. 400 BCE). She not only acquired an exceptional education but influenced politics as well. As a "foreign" resident in Athens, she had greater freedom than Athenian women did. At the insistence of her father, she was educated by her mother and by household slaves. (In ancient Greece some slaves were used as educators, though most were not educated.) As a young woman, Aspasia moved to Athens and became the consort of Pericles, a highly respected and powerful statesman and general. Aspasia was one of the few women to socialize with the leading thinkers of the time. She was well respected for her skills in rhetoric, and her abilities were acknowledged by several prominent Greeks, including Plato. She has been credited with ghost-writing speeches for Pericles, including his popular Funeral Oration.

However, even talents such as those possessed by Aspasia would never be enough to earn a woman equal political status with even the lowest class of male Athenian citizens. Given women's place in society during Plato's lifetime, his conception of an ideal state where women could rule stand outs as quite radical. Ideas similar to his regarding the education of women were in the air, and even parodied by Aristophanes in his comedies, but they were not widely held or discussed by the general public (Plato 1974, p. 225).

PLATO'S ARGUMENT

Before looking at Plato's work directly, it needs to be pointed out that there is some controversy among scholars about his views on women. His work has been charged with misogyny, on the one hand, and heralded as radically feminist on the other (see Bluestone 1987, Tuana 1994). It has been suggested by some scholars that the two extremes can be better reconciled if we distinguish between how Plato perceived Athenian women, and what he saw as women's natural abilities. Most scholars agree that Plato's view of Athenian women was not feminist, insofar as he was clearly opposed to equality for women as a group, just as he was opposed to democracy in general and to equality between men (Plato 1974, p. 563). He also chastised certain ideas as stemming from a "womanish and small intellect" (Plato 1974, p. 469). This accounts for at least some of the claims that he was anti-feminist. By contrast, in *The Republic* Plato outlines the ideal state and sees women as having the same capacities as men, and thus capable of sharing in any possible profession, if they possess the talent for it, and have been provided with the education and training needed. It is also thought that two women attended Plato's Academy and were taught by him.

With the preceding qualification, we can now look at Plato's *Republic*, which, like most of his writings, is in the form of a dialogue. Socrates, Plato's teacher, is presented as the protagonist and puts forward the arguments for the ideal state to several acquaintances. Through the figure of Socrates Plato envisions two groups: guardians and workers. The guardians are further divided into rulers and their auxiliaries. Family and private property are abolished among the guardians, so that none of them has any private interests, and all can maintain objectivity and deal fairly with the needs of the community. Plato's vision is also a eugenic one: women and men of similar capacities should marry and have children, in effect breeding like horses to get the best offspring. If capacities should prove wanting in a child of a higher group, or be present in a child of a lower group, the child can be moved between the groups. Plato also speaks of "quietly and secretly disposing" of "defective offspring," though it is not clear if he meant infanticide or removal from the community (Plato 1974, p. 461).

Although Plato's vision is not egalitarian, it does allow for capacity or potential to be found anywhere, regardless of sex or social status. Drawing an analogy with watchdogs, in which female watchdogs are employed to guard just as male ones are, Plato argues that women, like men, should share all duties. He makes three important claims to support this argument.

First, Plato argues that the natural differences between men and women, once the term "different" is clarified, are irrelevant to employment in society. The most obvious difference between men and women is the latter's capacity to bear children (Plato 1974, p. 454). In Plato's view this capacity is not relevant to a woman's natural abilities or talents for a certain occupation. A man and a woman may share the same nature regarding medical ability, for example, in spite of having different natures (Plato 1974, p. 454). As for child-rearing, Plato relegates it to workers, male and female, so that female guardians can have both the time and the energy to do their jobs. His reply to the comment that "child-bearing will be an easy job for the guardians' wives on those conditions" is striking: "which is as it should be" (Plato 1974, p. 460). Thus, for Plato, women's natural capacity to bear children does not carry with it the obligation to take care of them. Moreover, males who are workers are considered to be as capable of child-rearing as their female counterparts. In Plato's view, then, biology is not destiny, for either sex.

Second, Plato argues that the natural abilities that women share with men include not only medicine, music, athletics and even soldiering, but also philosophy (Plato 1974, p. 456). Janet Farrell Smith (1983, pp. 31–32) argues that this is Plato's "burden of proof argument," rather than a definitive claim that women have the same capacities as men. In her view, Plato simply puts the burden of proof on those who oppose women's education to show that women ought not to be educated. In this way, she argues, he leaves open the question of women's capacities. Since philosophy is required for the guardians in Plato's ideal state, and women may practise it, Plato concludes that women can be guardians.

Plato does comment, without argument, that women are weaker than men, presumably possessing their abilities to a lesser degree. Although there is much literature about this, the best reading seems to be that Plato regarded women, as a group, as being weaker than men, but without doubt allowed for the superiority of individual women over individual men (Smith 1983, p. 36). The salient difference is between men and women of similar natures, not between every woman and any man. For example, a woman may be superior to almost all men, as would be the case for a female guardian. Following this view, Plato calls for his male guardians to be matched with women who have "nearly similar natural capacities" (Plato 1974, p. 458).

Third, Plato claims that "if we are going to use men and women for the same purposes, we must teach them the same things" (Plato 1974, p. 451) In order to become rulers or to work as auxiliaries, must have the same access to education and training as men. Plato's comment on the Athenian practice of denying women the same intellectual and physical training as men is that it is "unnatural." Plato's guiding principle is that those with the same natural capacities ought to share the same occupations. Access to the same occupations requires access to the same training (Plato 1974, p. 456). According to Plato, any society that denies this is going against nature.

So we see that Plato provides a rational argument, based on natural capacity, as to why women ought to be afforded equal access to education and professions. One's biology, and in particular women's child-bearing capacity and men's lack of it, does not carry with it an obligation to take care of children, nor, in the case of men, any incapacity to take care of them.

Being aware of Plato's arguments for the education of a select group of women, and their equal participation in society, shows us that plausible arguments for the equality of elite women, though not of all women, have been available for over two thousand years. The arguments put forward by Plato, and by others in later periods, have been replied to by a few, but they have mostly been ignored. It seems, then, that resistance to women's equality is not because of the absence of rational arguments.

Familiarity with Plato's arguments is important, nonetheless, because they introduce an important theme around which some feminist and anti-feminist debates turn. The idea is that equality and inequality are somehow related to ability, capacity, or potential. Plato's argument for allowing certain women to participate equally in society clearly exemplifies this view. The deeper principle, as mentioned, is that a society's structure ought to be based on what is "natural."

There are two counterarguments against Plato that follow the principle that society ought to be based on what is natural. The first is that a woman's natural ability to bear children carries with it a "natural" ability and obligation to care for them. The second counterargument is simply the denial that women have the same capacities or abilities as men when it comes to intellectual and professional endeavours. The reliance on "nature" is the same, but where each argument differs is in defining what "nature" is, and what it may or may not imply about abilities and obligations.

At its inception in the 19th century, first-wave feminism did not address women's unequal access to public roles so much as women's right to receive education because they possess rational minds. Plato, in contrast, radically envisions a transformed society in which a woman may fill any public role that a man might, and assume a role in which she is not tied to the domestic sphere any more than a man would be. Plato envisions women becoming just like men, a view that was radical for his time, and a view for which liberal feminists are sometimes criticized today. In both cases, it is claimed that simply arguing that women are equally capable of filling any role that men are leaves the structure of society itself, and its values, hierarchies, and institutions, unchallenged. Again, equality for elite women is not the same as equality for all women.

Plato's *Republic* also touches on the interplay between a person's natural capacities and his or her socialization or schooling, which today is called "nurture." For Plato, natural capacity requires proper training and education, and this indeed is what his argument for the education of girls and women rests on. The debate over whether women and men are naturally different, or only different because they are nurtured or socialized differently, still rages, as do views about what to do with these alleged differences (to be discussed in a later section).

There are arguments against Plato's position, largely based on denying that women have the same capacities as men or on the claim that child-bearing capacity carries with it child-rearing responsibilities. However, one could argue that everyone, regardless of natural capacity, has a right to education. Plato's discussion of education is a good example of one view of the interplay between nature on one side and nurture on the other. For Plato, natural capacities may be found anywhere, but still require nurturing in order to be cultivated.

ARISTOTLE'S VIEW

Aristotle (384–322 BCE) argued that "the male is by nature fitter for command than the female, just as the elder and full-grown is superior to the younger and more immature" (Bell 1983, p. 66). While Plato thought that women shared the same capacities or virtues as men, Aristotle saw such virtues as different in kind, so that "the courage and justice of a man and of a woman are not the same; the courage of a man is shown in commanding, of a woman in obeying. And this holds of all other virtues" (Bell 1983, p. 68). There are several other passages in which Aristotle explains the inferiority of women. In men, "qualities or capacities are found in their perfection, whereas women are less balanced, more easily moved to tears . . . more jealous; she is also more false of speech [and] more deceptive" (Bell 1983, p. 66). A woman may rule, but only as an heiress, and therefore not in virtue of her own excellence but due to wealth or power. Elsewhere, Aristotle writes of women possessing the same virtues as men, but to a lesser degree (Bell 1983, pp. 65–66).

Like Plato, Aristotle is among the most influential philosophers in the western tradition. Aristotle was considerably more prolific, however, and much more of his writing has survived the centuries, with the result that, in some respects, his influence has exceeded Plato's. Aristotle provided substantial and systematic treatments of metaphysics, physics, and biology, contributing to the formation of the metaphysical and scientific worldview that dominated western thinking until the 19th century. (For example, what is now known as the Aristotelian–Ptolemaic worldview held that the Earth was the centre of the universe. This view was overturned and eventually replaced by Copernicus, Kepler, and others.) Aristotle's view of women is woven in with his biological and metaphysical views. He argued that "just as it sometimes happens that deformed offspring are produced by deformed parents, and sometimes not, so the offspring produced by a female are sometimes female, sometimes male," because "the female is, as it were, a deformed male" (Bell 1983, p. 63). In his *Generation of Animals* (1953, quoted in Bell 1983, p. 63), he describes procreation as follows:

> The menstrual discharge is semen, though in an impure condition; i.e., it lacks one constituent, and only one, the Principle of Soul. . . . An animal is a living body, a body with Soul in it. The female always provides the material, the male provides that which fashions the material into shape. . . . Thus, the physical part, the body, comes from the female, and the Soul from the male, since the Soul is the essence of a particular body.

In this biological explanation of the difference between men and women, centred on their respective roles in procreation, Aristotle follows his general metaphysical view that everything consists of both form and matter. A chair's structure, for example, is its form, while its matter is the wood (Aristotle 1993, p. 412). In procreation, women provide the material and men provide the form, or Soul, and the result is a living body with a Soul in it. Aristotle did not mean "soul" in the religious sense as we understand it today, but his view fits well with Christianity insofar as he divided everything up into matter and form, and form can be equated with "soul" in the Christian sense. (More will be said later about Aristotle's views and his influence on other thinkers.)

This is one way in which women came to be associated with the material world, the body, and passivity, and men with the more abstract immaterial world and with action. Earlier philosophers had also categorized the male and the female with sets of attributes and associations, the male ones being considered more desirable. For example, the followers of Pythagoras (mid-6th century BCE)—a philosopher whose very existence has been questioned, but whose movement, or "school," certainly influenced both Plato and Aristotle— constructed a "table of opposites" in which "male" belonged to the category that also contained "limited" (or definite), "oddness" (of numbers), "unity," "right," "rest," "straight," "light," "good," and "square," while "female" was linked with "plurality," "left," "movement," "the curved," "dark," "bad," and "oblong." Of note is that, unlike Aristotle, the Pythagoreans associated movement, not rest, with the female. It should also be noted that women were said to have attended Pythagoras' school.

FROM ARISTOTLE TO AQUINAS

Today, when we think of science we usually think of it as separate from, and sometimes even antithetical to, religion.

In the Middle Ages, in contrast, science and religion were not distinct entities, and what we call science was regarded as "natural philosophy." It was not until the 17th century that science began to occupy a separate niche and develop distinct methods of inquiry, but even then it was still very much mixed in with theology, alchemy, and magic.

In the Middle Ages, the views of natural philosophers such as Aristotle were woven into theological works. Aristotle's philosophy, along with his remarks about women, were reintroduced into European intellectual history in the 12th century through Syrian and other Arabian scholars, notably Avicenna (c. 980–1037 CE) and Averroes (1126–1198). The commentaries of these philosophers began to reach Europe through Spain and Italy, and in particular through Thomas Aquinas (c. 1225–1274), an Italian Dominican who sought to reconcile Aristotle's works with Christian doctrine. Aquinas had several of Aristotle's works translated, including the *Generation of Animals*.

David F. Noble suggests that the medical and scientific ideas of the ancient world were appropriated in support of the misogyny of the medieval religious world (Noble 1992, p. 157) The misogyny in early religious writings is sometimes strikingly overt. Religious writers appealed to religion to justify or reaffirm social structures in which women had few rights. One example is the Church Father Tertullian (c.160–c.220 CE), who described women as heirs to the curse of Eve, "the Devil's gateway, . . . the unsealer of that (forbidden) tree, . . . [and] the first deserter of the divine." Reacting to women who had been enjoying participation in the public realm by teaching and speaking in church, he stated that women, inferior in nature and carrying the guilt of original sin, ought to be humble and silent, not speak in church, and not question their husbands.

Some women suffered at the hands of Christianity in a different way. Hypatia (c. 370–415), a mathematician, astronomer, and Neo-Platonist philosopher, lived in Alexandria in Egypt. As with Aspasia, it was a matter of luck that Hypatia's exceptional talents were given a chance to prosper through education. Her father taught her mathematics and philosophy, and she not only became the head of her father's Platonist school, but became renowned for her abilities and her charismatic teaching. Her social position was brought into prominence because she was a woman, and also because she was viewed as a pagan and not a Christian, given her philosophical views. Caught in the middle of a battle for power between church and state in Alexandria, she was considered as a serious threat by the Christians because of her scholarship, the depth of her knowledge, and her influence. This led to her brutal public murder at the hands of Christians, as described by the 5th-century historian Socrates Scholasticus (cited in Alic 1986, pp. 45–46):

> They pull her out of her chariot: they haul her into the church called Caesarium: they stripped her stark naked: they raze the skin and rend the flesh of her body with sharp shells, until the breath departed out of her body: they quarter her body: they bring her quarters unto a place called Cinaron and burn them to ashes.

By the Middle Ages the misogyny in Christianity was clearly entrenched. Scholars who were working on reconciling the philosophy of the ancient world with Christianity also incorporated ancient views of women. Scholars had access to both Plato and Aristotle, but, as we shall see, Aristotle's view eventually prevailed. Nevertheless, although it might seem that views such as Aristotle's might easily lend themselves to misogynist Christian doctrine, some work was required by scholars to smooth over inconsistencies.

One of the problems that had to be reconciled by Aquinas when considering Aristotle's view of women was the tension between his theory that women were deformed men and the doctrine that they were created by God in His own image. In his *Summa Theologica* Aquinas imagines the following objection (quoted in Bell 1983, p. 102):

> It would seem that woman should not have been made in the first production of things. For the Philosopher [Aristotle] says, that the *female is a misbegotten male*. But nothing misbegotten or defective should have been in the first production of things. Therefore woman should not have been made at that first production.

The problem for Aquinas was both scientific and theological: how could a perfect God have created something imperfect? Aquinas' response (as quoted in Bell 1983, p. 103) is as follows:

> As regards the individual nature, woman is defective and misbegotten, for the active force in the male seed tends to the production of perfect likeness in the masculine sex; while the production of woman comes from a defect in the active force or from some material indisposition, or even from some external influence. . . . On the other hand, as regards human nature in general, woman is not misbegotten, but is included in nature's intention as directed to the work of generation. Now the general intention of nature depends on God, Who is the universal Author of nature. Therefore, in producing nature, God formed not only the male but also the female.

Thus Aquinas adopted Aristotle's opinion that a male child is perfect and a female child is the result of a defect, as an explanation of what was taken to be fact, namely, the inferiority of women, and supplemented it with the claim that a woman is not misbegotten insofar as she fulfils what is required for the human species to reproduce.

AVERROES, PLATO, AND WOMEN

Other approaches were available to medieval theologians and philosophers in addition to Aristotelianism. The Muslim philosopher, theologian, and scientist Averroes, who, as mentioned, was partly responsible for the rediscovery of Aristotle's works, chose not to adopt Aristotle's views on everything. Like Aquinas, he too sought to synthesize what he knew of the ancient Greek tradition with religion. However, although he admired Aristotle and translated several of his works, Averroes also read Plato's *Republic*, and espoused more of the Platonic view regarding the equality of men and women, and their rights and duties as full citizens. Like Plato, he saw men and women differing only in degree, but he also saw each as being better at some things than at others. In the spirit of *The Republic*, Averroes commented on the effect of society on women (Averroes 1974, p. 59):

> Our society allows no scope for the development of women's talents. They seem to be destined exclusively to childbirth and the care of children, and this state of servility has destroyed their capacity for larger matters. It is thus that we see no women endowed with moral virtues; they live their lives like vegetables, devoting themselves to their husbands. From this stems the misery that pervades our cities, for women outnumber men by more than double and cannot procure the necessities of life by their own labours.

It is remarkable that such words were written in the 12th century. What we see in Averroes' comment are the Platonic ideas that women share talents with men, that they are not destined exclusively to bear and raise children, and that their talents need to be cultivated. Pressing beyond Plato, however, Averroes adds the further claim that women ought to be able to support themselves. Also present are the ideas, which we see again in the 17th century that allowing women more opportunities will improve their moral virtue.

Unlike Aquinas's views, Averroes' arguments for increasing women's freedom did not prevail. Although Averroes was highly influential, and his work was translated into Latin in the 13th century, he often caused controversy in the West and was finally condemned as heretical in 1227, though not simply because of his views of women.

PARACELSUS AND WOMEN

The Swiss-born polymath Auroleus Philippus Theostratus Bombastus von Hohenheim (1493–1541) has been immortalized under his more manageable pen name, "Paracelsus." The son of a physician, Paracelsus worked within the hermetic tradition, focusing on medicine and alchemy, and was noted for his hermaphroditic view of the universe. In zoology and botany, a hermaphrodite is an organism that possesses both male and female sex organs. His model contrasted with Aristotle's. Paracelsus writes (1951, p. 101),

> When the seed is received in the womb, nature combines the seed of the man and the seed of the woman. Of the two seeds the better and the stronger will form the other according to its nature. . . . The seed from the man's brain and that from the woman's brain together make only one brain; but the child's brain is formed according to the one which is the stronger of the two, and it becomes like this seed but never completely like it.

In this biological explanation, unlike Aristotle's, the birth of a female child is the result of a stronger seed, as opposed to a defective one. Paracelsus often rails against women in his writings, but he nevertheless has more of an egalitarian explanation of procreation. This may have stemmed partly from the hermetic tradition, in which nature was viewed in terms of both male and female principles in a hermaphroditic union (Keller 1985, p. 49).

ARISTOTLE PREVAILS

It was Aquinas's ideas about women that won out over Averroes', or, in effect, Aristotle's over Plato's, and it was the mechanistic and dualistic science of the 17th century that won out over the hermetic tradition represented by Paracelsus (Keller 1985, p. 54). Perhaps it is more accurate to say that Aristotle's and Aquinas' views both reflected and helped to perpetuate views that were already patriarchal in form and well entrenched. Plato, Averroes, and Paracelsus offered alternatives, but theirs were small voices in a large crowd. Importantly, then, we can see that arguments for the equality of women did exist, but were ignored. We also see that, in addition to ignoring arguments for the equality of women, scholars have, in some cases, actively defended the patriarchal views exemplified in Aquinas's work.

Aristotle described women as being different from men in two ways: women are deficient, or misbegotten males; women are somehow distinct from and inferior to males, but in a complementary fashion. It is generally agreed that the ancient Greeks subscribed to the view that females were lesser males. This view can be extracted from the writings of both Plato and Aristotle, but there are some subtle differences as well. Aristotle also claims that, although we use the same name to describe the virtues of men and those of women, these virtues manifest themselves differently, in a complementary fashion. A man's courage, for instance, is exemplified by commanding, a woman's by obeying. For Plato, women as a group are lesser males, but, in stark contrast to Aristotle, he argues that individual women may possess greater talents than some or even most men. As with Aristotle and others following him, women's inferiority was taken as a given and established fact, as opposed to something determined by proper scientific study or rational thought. This is an example of bias infecting philosophical discussions, scientific investigation, and interpretation of results. For example, women's inferiority, in the view of the craniologists of the 19th century, was a fact to be explained, not one to be determined.

It is important to note that the view of women as being different from men is not held solely by anti-feminists. Some feminists, both male and female, have also argued that women have different capacities, viewpoints, or cognitive abilities that complement men's. These views also divide along the lines of nature and nurture, some arguing that women by nature are different, and thus should be assigned different roles, while others argue that women's different social experiences afford them different vantage points from which to see the world and provide enriching perspectives in their fields of endeavour.

⟋⟍∞⟍⟍

CHAPTER 2

Renaissance and Enlightenment

In the early modern era (15th to 18th centuries), a few exceptional men and women argued in favour of women's abilities and for the provision of universal access to education, and some even fought for women's right to hold public positions. However, the majority of thinkers continued to put forward Aristotle's and Augustine's vision of women, supporting severe limits on the access of girls and women to education. In general, these thinkers expressed views that fell within one or other of the two categories already discussed: either women are deficient males or they are inherently different from them. Although some of these views may seem to be complimentary to women, the ultimate conclusion that these thinkers generally reached was that women cannot, or should not, participate in the higher forms of thinking required for philosophical and scientific endeavour. It was views such as these that, for a very long period, prevented women from achieving equal access to education and to science. Even when such social barriers were removed, women still had to overcome psychological barriers, such as feelings of insecurity and low levels of self-confidence, that stemmed from over two thousand years of intellectual history that deemed them, for the most part, unintelligent and incapable of higher forms of thought. This chapter presents examples to demonstrate the type of arguments used during this era on both sides, for and against women, and the general attitudes towards women held by both sexes.

FRANÇOIS POULLAIN DE LA BARRE

In the first part of his *Méditations métaphysiques* (1641) the French philosopher René Descartes famously and influentially brought reason and the individual to the forefront through his use of introspection and the method of doubt. Using reason alone to establish what knowledge is certain, he meditated on what cannot be doubted. The indubitable starting point he discovered was "I think, therefore I am," for to doubt is to think and to think is to be. Some commentators have suggested that by detaching the mind from the sensible world (our bodies) and extending the potential to reason to everyone, Descartes' work opened up a space for women (Perry 1986, p. 70; Perry 1999, p. 184; Waters 2000, p. 67; Schiebinger 1999, p. 111).

In *De l'Égalité des deux Sexes* (1672, *The Equality of the Two Sexes*) François Poullain de la Barre (1647–1725) extended Descartes' views in just this way. Poullain argues (as quoted in Lewison 1989, p. 85) that

> it is easy to realize that the difference between the sexes concerns only the body—being, correctly, only [in] this part that serves for the production of men. Since the mind participates [in this activity] only by giving its assent (and giving it in all people in the same manner), we can conclude that it is sexless.

Poullain's point is that men and women differ physically in terms of reproduction but not in terms of the mind. The body is simply for reproduction, and the mind, which participates in reproduction only by giving its assent, is sexless. This assertion was in itself a radical departure from the mainstream view, established by Aristotle and his later interpreters, that women's capacity to bear children is directly connected with their inferiority. If the mind is sexless, as Poullain claims, then women's intellectual potential is no different from men's.

Poullain goes on to argue that the intellectual capacities of men and women are indeed equal. Invoking religion much as Aquinas did, he writes (as quoted in Lewison 1989, p. 85),

> God joins the mind to the flesh of a woman as to that of a man, and He unites them by the same laws. Feelings, passions, and will make and maintain this union. And the mind, not functioning differently in one sex than in the other, is equally capable of the same things [in both].

The mind is joined to the body by the same laws and interacts in the same way, through the feelings, passions, and the will. The only difference between men and women, as Poullain sees it, is in the form of the flesh to which the sexless mind is attached. By separating the mind from the body, one can account for the differences between men and women, and at the same time argue for their mental equality.

This is one way to explain the difference and the sameness between the sexes, but it is not necessarily a knockdown argument against those who hold that women have minds, but *inferior* ones or *different* ones. One might maintain, for example, that women possess some level of reasoning ability, but still adopt Immanuel Kant's view that women do not possess deep understanding, or William James's view that they are incapable of abstract thinking, or David Hume's view that they are too easily swayed by emotion and hence not governed by reason. The point is that neither argument, for or against, is definitive. There is always wiggle room.

As we have seen, the arguments claiming that women have inferior minds do not take proper account of the existence of intellectually accomplished women, who are conveniently dismissed as being anomalies or exceptions. These positions also ignore the ways in which women have been socialized and educated, and assume that a woman's ignorance is a reflection of her nature. Given these considerations, one may find that the view that women have inferior minds does not offer a good explanation of why there are, and have always been, intellectually gifted women, and also fails to account for the role of society and education in a person's development and growth. It is perhaps for these reasons that Poullain bolstered his dualist argument by pointing out how the structure of society favoured men, as it is men who create the laws. Like John Stuart Mill in the 19th century, Poullain highlights the role of society and points out the error of attributing what has been cultivated in women to women's natures. He writes (as quoted in Lewison 1994, p. 85),

> We must take into account that those who have made or compiled the laws, being men, have favoured their sex (as women would perhaps have done if they had been in their place). And the laws having been laid down from the beginning of society as they are now with respect to women, legal scholars, who also have their prejudices, have attributed to nature a distinction that derives from custom alone.

According to Poullain, those who think that women are inherently inferior have confused a woman's nature with a woman's upbringing and the society she lives in, that is, custom. Although the mind has no sex, a person who is treated differently by society and placed in a different role by custom will appear, misleadingly, to have a different nature.

NICOLAS MALEBRANCHE

Unfortunately, Poullain's arguments made little headway against the conventional wisdom of his day, represented by such statements as that of David Abercromby, a Scottish physician and writer, in his treatise *A Discourse of Wit* (1685), that women "have not received from God so perfect Souls as Men, because by God's special appointment they are to obey and Men to command" (Abercromby 2003, pp. 214–15). Such views, based on ideas inherited by Aristotle, were bolstered by the attitudes of the churches. One of the strongest voices raised in the 17th century against education for women was that of the French Catholic priest Nicolas Malebranche (1638–1715), who, like Poullain, was a Cartesian philosopher, but came to very different conclusions.

Malebranche endorsed the commonly accepted belief that men were superior to women. Couching his arguments in terms then current, that is, invoking nature and women's anatomy, he described what he saw as the inadequacy of the female mind. Malebranche was ready to accept that there were some women of "sterling mental ability," but he insisted that in general women are preoccupied with taste, fashion, etiquette, and dress. Of course, this would have been true of many women in Malebranche's time, since their role was confined to finding a husband, and opportunities for receiving an education comparable to men's was almost unheard of. It seems, then, that Malebranche's superficial assessment of women's abilities did not take any account of the constraints placed on them (Phillips 1990, pp. 14–15).

One anonymous female writer presented a direct counter-argument to Malebranche in an essay in the *Lady's Journal*, published by Peter Motteux (1660–1718), an English journalist, translator, and playwright of French origin (he had been christened Pierre). Motteux regularly published *Gentlemen's Journal* between January 1692 and October/ November 1694. The single volume of the *Lady's Journal*, published in October 1693, was the first magazine written for and by women, and provided a platform for the side of the debate that supported women. Patricia Phillips (1990, p. 20) summarizes the most important points made by the anonymous writer: that "women may apply themselves to the Liberal Arts and Sciences," and "that women be excluded no longer from education that men enjoy, so that they too may be admitted to the 'honourable and profitable employment' which would then be open to them, this being 'the chief end of most men's study."

MARGARET CAVENDISH

However, even prominent and educated women, such as Margaret Cavendish, Duchess of Newcastle-upon-Tyne (1623–73), adhered to the belief in female inferiority. In her book *The World's Olio* (1655) Cavendish refers to women's brains as "cold, moist, and slow," and writes (Cavendish 2000, p. 138) that "Women have no strength nor light of Understanding, but what is given them from Men; this is the Reason why we are not Mathematicians, Arithmeticians, Logicians, Geometricians, Cosmographers, and the like." This statement seems strange, coming as it does from a woman who defied the conventions of the day by writing several books on science and philosophy. Londa Schiebinger (1989, p. 58) points out how surprising it now seems that Cavendish did not blame lack of education for what she saw as women's inferiority, summarizes the contrast between Cavendish's writings and her own way of life:

> Cavendish's hesitant approach to the woman question was never consistent with her own ambitions. She refused from her earliest years to follow a traditionally female path. In her youth, she took up the pen and not the needle. In her maturity, she took up philosophy and not housewifery . . . [and] her voluminous publication, her visit to the Royal

Society, her autobiography, her early atheism, her criticism
of "learned men" overstepped the bounds of convention.

Perhaps Cavendish could have encouraged other women
to seek the kind of education she herself had received if she
had achieved more consistency about who she was and what
she wrote. In each historical period there have been women
and men supportive of women's education, who believed that
women could learn as well as men.

LOCKE AND LEIBNIZ

Second only to Descartes in his influence on western
philosophy, politics and society, the English thinker John
Locke (1632–1704) argued for individual rights based on
equality and grounded in nature. In his view, it is because
human beings are "furnished with like faculties" that they
are equal (Locke 1980, p. 6). Locke appealed to what he
saw as the natural state of man, a state of perfect freedom, to
argue against the absolute power of monarchy and in favour
of individual rights: men by nature are "all free, equal, and
independent," and "no one can be put out off his estate, and
be subjected to the political power of another, without his
own consent" (Locke 1980, p. 95). Freedom is governed only
by certain laws of nature, recognized by reason, that limit
one's actions to those that do not invade the rights of others
(Locke 1980, pp. 6–7 and 63). According to Locke, we are
free in our natural state to make a conscious choice to join
together into a society, a "body politic" or "commonwealth"
that is run by the majority (Locke 1980, p. 95). The chief
reason for entering into a society, as opposed to living in
absolute freedom, is to preserve one's private property (Locke
1980, p. 127) In this way, we all remain free but agree, for
our own benefit, to follow certain rules, so that we can live as
peacefully as possible and without anxiety that our property
will be invaded.

However, although Locke supported a government chosen by the people, as opposed to monarchy, and argued for equal rights among men, he did not extend those rights to women, or to men who did not own property. In his influential conception of private property, he argued that it is a man's labour that gives him his right to own private property, as opposed to his ascribed social status. He did not argue for the right of women to own property (Locke 1980, pp. 30 and 32). Similarly, although he argued against the power of monarchs over their subjects, he did not argue against the power of husbands over their wives. Locke also excluded the labouring masses from the freedom and equality he envisioned, on the basis of the inferior rationality that they shared with women (see Eisenstein 1981). Thus, as with early Athenian democracy, when one asks "Equal rights for whom?" the answer is "Not for everyone."

Locke did believe that boys and girls should be educated in a like manner, but he wrote little about higher education and did not envision adult women as inhabiting anything more than the private, domestic sphere (Eisenstein 1981, p. 48) He was an advocate of home schooling and thought that women's education was imperative in order to ensure adequate home schooling for their children, which is, of course, a very different matter from advocating education for women in their own right (O'Day 1982, p. 188).

Because of these arguments for the separate roles of men and women, Locke is sometimes seen as being partly responsible for splitting what he called political or civil society from the family, by calling for equality among men in the civil or public realm, while maintaining the rule of husbands over wives in the private realm (Eisenstein 1981, pp. 33, 42–43, and 47–48) This reflects another important distinction in feminist scholarship, namely, the split between the public and the private. Women were relegated to the private realm and kept out of the all-male public realm, and each was governed by a different set of rules.

Ironically, it was this very difference between the public and private realms that was cited by Locke's contemporary, the German philosopher and mathematician Gottfried Wilhelm von Leibniz (1646–1716), in his argument *in favour* of education for at least a minority of talented women (as quoted in Schiebinger 1989, pp. 39–41):

> I have often thought that women of elevated mind advance knowledge more properly than do men. Men, taken up by their affairs, often care no more than necessary about knowledge; women, whose condition puts them above troublesome and laborious cares, are more detached, and therefore more capable of contemplating the good and beautiful.

It is noteworthy, however, that Leibniz put this argument forward in a private letter to a royal patron, rather than in the public realm itself, where Locke's conclusions went largely unchallenged for generations.

MARY ASTELL

However, Locke's conclusions did meet with some limited criticism, notably from Mary Astell (1668–1731), who argued in the early 18th century in favour of an institution that was not to be created until the late 19th century: colleges for women. Astell's view was similar to Plato's, in seeing education as a crucial part of fostering the equal talents women possessed (in Waters 2000, pp. 40 and 42):

> The Incapacity [of women], if there be any, is acquired and not natural . . . Women are from their Infancy debar'd those Advantages, for the want of which they are afterwards reproached . . . For since GOD has given Women as well as Men intelligent Souls, why should they be forbidden to improve them? Since he has not denied us the faculty of

Thinking, why shou'd we not (at least in gratitude to him)
employ our Thoughts on himself their noblest Object . . .
and not unworthily bestow them on Trifles and Gaities and
secular Affairs?

Astell extended Locke's principles in order to secure certain
rights for women, in particular, the right to education. Like
Locke, Astell sees education from an early age as crucial,
but for Astell the context within which education takes place
is as important as the particular subjects taught. The right
to education does not start at college, nor does it simply
end when schooling ends. Like Poullain, Astell (in Waters
2000, p. 49) points to social customs that afford men greater
advantages, both from birth and after they have been educated.
Girls receive little or no education, while boys are

early initiated in the Sciences, are made acquainted with
Ancient and Modern Discoveries, they Study Books of
Men, have all imaginable encouragement; not only fame,
a dry Reward nowadays, but also Title, Authority, Power,
and Riches themselves, which purchase all things, are the
Reward for their Improvement.

As will be seen in a later chapter, double standards that
favour men over women for awards, prizes, promotions,
and recognition still dominate the culture of science in the
20th century. What Astell identifies in her radical writings is a
whole social structure set up to encourage boys but not girls.
In her view, it is not just access to training and education that
is important but also the environment in which these activities
take place. Astell thus deepens the analysis of what is meant
by nurturing, socializing, and bias. The fault of women, if
there is one, lies not in their nature, but in their nurture and
in the advantages they have been denied. What Astell also
provides is a different explanation of the observed fact that
some women were inferior or frivolous. Whereas most thinkers

of the time assumed that this attitude was based on women's nature, Astell saw it as having been cultivated by society, or, as we might put it nowadays, socially constructed. (Mary Astell's life and work are discussed further in Chapter 5.)

JEAN-JACQUES ROUSSEAU

While Mary Astell was effectively writing before her time, Jean-Jacques Rousseau (1712–1778) was perfectly in tune with his, at least on the question of women, their nature and their capacities. Rousseau famously believed that men are born equal, and become unequal due to the structure of society and its institutions, yet he failed to apply this insight to women. Instead, he argued that women's natures are essentially different and that they possess virtues (or capacities) different from those of men. In his book *Émile, ou, de l'éducation* (first published in 1762), Rousseau (1911, p. 442) describes the differences between men and women, and their gender roles:

> In the union of the sexes each alike contributes to the common end, but in different ways. From this diversity springs the first difference which may be observed between man and woman in their moral relations. The man should be strong and active; the woman should be weak and passive; the one must have both the power and the will; it is enough that the other should offer little resistance. When this principle is admitted, it follows that woman is specially made for man's delight. . . . I grant you this is not the law of love, but it is the law of nature.

For Rousseau (1911, p. 480) it follows that

> the search for abstract and speculative truths, for principles and axioms in science, for all that tends to wide generalisation, is beyond a woman's grasp; their studies should be thoroughly practical. . . . A woman's thoughts, beyond the

range of her immediate duties, should be directed to the study of men, or the acquirement of that agreeable learning whose sole end is the formation of taste.

On this question Rousseau was far from being as radical as his reputation may suggest. Indeed, in ascribing fixed but complementary natures to men and women, he reproduces the tradition handed down from Aristotle, who wrote that "the courage of a man is shown in commanding, of a woman in obeying" (quoted in Bell 1983, 67–68). Rousseau also criticized Plato for writing, in *The Republic*, that men and women could hold the same public positions if they had received the same education. He argues vehemently against making the two sexes equal, for, if women were educated like men, they would stop controlling men with their charm, femininity, and dependence. Rousseau maintains that a woman is worth more as a woman than she would be as a man, and that her education must be conducted accordingly (Rousseau 1911, p. 451):

A woman's education must therefore be planned in relation to man. To be pleasing in his sight, to win his respect and love, to train him in childhood, to tend him in manhood, to counsel and console, to make his life pleasant and happy, these are the duties of woman for all time, and this is what she should be taught while she is young.

For Rousseau, then, biology is destiny, and women's lives have to be entirely focused on, and subjected to, men's needs and pleasures. Women are not to be independent and should receive only a limited education, directly linked to their roles as nurturing wives and mothers.

HUME, KANT, GOETHE, AND HUMBOLDT

Before Rousseau's ideas could have influenced him, the Scottish philosopher David Hume (1711–1776) had amicably remarked (as quoted in Bell 1983, p. 157) that, while "women of sense and education . . . are much better judges of all polite writing than men of the same degree of understanding," women have a "great share of the tender and amorous disposition, and therefore are too easily swayed by their emotions." In fact, most philosophers and educators in the 18th century, and even most educated women, embraced an approach similar to Rousseau's and Hume's, and perpetuated similar ideas, not necessarily because they had read the works of cither philosopher, but because Rousseau and Hume had expressed the conventional wisdom of the era.

Thus, for example, the German philosopher Immanuel Kant (1724–1804), in his early work *Beobachtungen über das Gefühl des Schönen und Erhabenen* (1764, *Observations on the Feeling of the Beautiful and the Sublime*), describes woman as having an "inborn feeling for all that is beautiful, elegant, and decorated," and asserts that she has "just as much understanding as the male, but it is a *beautiful understanding*, whereas ours [men's] should be a *deep understanding*, an expression that signifies with the sublime." In his much later work *Anthropologie in pragmatischer Hinsicht* (1798, *Anthropology from a Pragmatic Point of View*), Kant draws the conclusion that "laborious learning or painful pondering, even if a woman should greatly succeed in it, [would] destroy the merits that are proper to her sex," and adds, "A woman who has a head full of Greek . . . or carries on fundamental controversies about mechanics . . . might as well even have a beard" (as quoted in Bell 1983, p. 242) For Kant as for Rousseau, men and women have different capacities and levels of understanding, and even if a woman somehow becomes capable of higher learning, it is highly undesirable that she should take part in it.

This is in stark contrast to Plato's view, as discussed in Chapter 1, but resonates with the assumptions and prejudices of several writers on the education of girls and boys in the 18th and 19th centuries. The German poet, novelist, playwright, and philosopher Johann Wolfgang von Goethe (1749–1832), made similar statements regarding women. As Phyllis Stock (1978, p. 108) points out, "Goethe wrote that even the most educated woman had more appetite than taste. We love everything in a woman except intelligence, [he wrote,] unless it is brilliant, or unless we already loved her before it was evident." Another German polymath, Wilhelm von Humboldt (1767–1835), asserted that men are more enlightened and women more emotional in his work "Über den Geschlechts-Unterschied" (1795, "On Sexual Difference"). In early modern Europe, few male thinkers cared to explore the distinction between nature and nurture, or to question the other ideas about women that they had inherited from their philosophical forebears, but continued to use arguments to maintain the status quo.

———✺———

CHAPTER 3

The Classic Arguments and Debates

This chapter explores some of the ways in which particular ideologies were used to support, or to challenge, the prevailing view that women are inferior to men, an argument used to justify the exclusion of women from higher education until only about a hundred years ago.

Change and improvement do not rest on rational argument alone. People are free to accept, reject, or simply ignore rational arguments, especially if they are already in positions of power and have a vested interest in maintaining the status quo. It is also possible for two rational arguments to have contradictory conclusions. In these cases, the deciding factor comes down to whether or not the premises of the arguments are true, and that in its turn is sometimes a very complicated and contentious matter. Examination of the premises of an argument can sometimes be blocked by prejudice and bias, but confusion over them can also be the result of people having different yet equally valuable points of view.

Thus, there is more to convincing people to adopt a point of view than simply presenting a rational argument. For example, we know that women did not win the right to vote by the sudden discovery and presentation of rational arguments alone. In fact, as we saw in Chapter 1, arguments for women's higher education and equal participation in society have been around since the 4th century BCE. At times, the women's movement was tied to social movements, such as the Civil War and the abolition of slavery in the United States, the emancipation of the lower classes in Britain, or the revolutions in Russia. These were times when ideas and social values were changing rapidly, not just as the result of rational arguments but also under the influence of changing sentiment, new experiences, and the undying perseverance of those committed to causes. Proponents of women's rights in North America and Europe worked tirelessly—lecturing, writing, holding conferences, and staging protests—in order to mobilize and persuade men and women alike that women ought to have equal rights with men.

We can say, then, that, while the ability to argue and to counterargue plays a role in mobilizing and persuading people, it is just one tool among many. Coupling this understanding with the fact that women's advances have been cyclical challenges the idea that misogynist views are simply "old-fashioned," and that it is a mark of our "progress" that we no longer, or to a lesser degree, treat women as second-class citizens, or as not being citizens at all. It is the belief in this myth that leads many to excuse misogynistic views as simply the products of a past era. Perhaps more importantly, belief in this myth also leads many to assume the converse: namely, that feminism is no longer needed and the gains that women have made are part of a natural continuum of progress that somehow takes care of itself. As Londa Schiebinger (1999, p. 32) has aptly put it,

> History dispels the myth of inevitable progress in respect to women in science. There is a sense that nature takes its course—that, given time, things right themselves.

The history of women in science, however, has not been characterized by a march of progress, but by cycles of advancement and retrenchment. Women's situation has changed along with social conditions and climates of opinion.

Belief in this myth is much more powerful and insidious than most people realize. It leads to a tendency to assume that, once gains have been won, they cannot be lost, and also to the assumption that new methods and forms of resistance to women's full participation in society, whether conscious or unconscious, cannot crop up. In fact, today many young people, men and women alike, believe that gender equality has been achieved, and that there are no more gender issues to be concerned about. Becoming familiar with the kinds of arguments that have been used to support the subjugation of women helps in recognizing and categorizing such arguments when they re-emerge in contemporary guise.

As we have seen, there have been recurring arguments, or sometimes just pseudo-arguments, supporting the view that women should *not* be given equal rights and opportunities. One point to note is that the denial of women's access to education and full participation in society has been actively fostered and supported, rather than just being complacently continued. Being aware of and familiar with the arguments puts one in a better position to understand how anti-feminist rhetoric is still being sustained today, and how deeply embedded it can be in any society, anywhere.

COMPLEMENT THEORIES AND BIOLOGY

The view that females are inherently distinct from males, yet complementary to them, is sometimes called "the natural complement theory" (Waters 2000, p. 67). According to this theory, women's different capacities suit them, by their very nature, to the domestic sphere and, in particular, to child-rearing. The theory can be used—and, of course, it has been—to justify the separation of the public and private realms along biological (sex) lines. It is generally agreed among scholars that this theory began to take shape in the 17th and 18th centuries, in contrast to the ancient Greek view; however, as we have seen, it is present to some degree in Aristotle's work.

There are contemporary forms of feminism that espouse a kind of complement theory, but the innate female capacities that they pick out do not tie women to the domestic sphere. Examples of such views are found in arguments that claim that women are contextual thinkers, complementing what is seen as the male abstract and detached form of thinking. The arguments vary as to the explanation they give for the difference: some say that it is a result of socialization (nurture), while others say that it is innate (nature). The same type of argument has occurred over the concept of race. Sex and race have been connected, in that women and blacks have both been argued to have relatively smaller brains and therefore less intelligence than whites and/or males, and have also been said to share other physical qualities that point to their supposed inferiority (see Stephans 1996).

For such arguments to work, however, one must also adhere to the principle that the structure of society should be based on people's natures. Plato used this idea, but in the opposite way, to justify women's right to participate in public roles and to be educated for these roles. According to complement theory, if talents or natures are divided along sexual lines, then society ought to reflect these differences by according different social roles to each sex. This view can then be used to justify pre-existing divisions of labour based on sex: Men control the public sphere (political and economic life), while women control the private sphere (domestic and family life).

Complement theories were often given biological explanations, which is why most of them fall under the heading of *natural* complement theories. Londa Schiebinger has pointed out that the number and influence of complement theories rose dramatically in the 17th and 18th centuries, at the same time as women began to be systematically excluded from science, which was increasingly formalized and removed from its older milieu of informal salons and artisan workshops (Schiebinger 1999, p. 111) She argues that with the infiltration into western society of Cartesian ideas in which the mind was considered sexless, new justifications were required to exclude women from intellectual endeavours. As we have seen in Chapter 2, complement theories did the job.

Evelyn Fox Keller (1985, p. 63) argues that during the 18th century, the period often referred to as the age of "scientific revolution," a variety of sociopolitical factors had resulted in a society with a heavily entrenched sex-based division of labour: "In sympathy with, and even in response to, the growing division between male and female, public and private, work and home, modern science opted for an ever greater polarization of mind and nature, reason and feeling, objective and subjective." Olwen Hufton, Leverhulme Research Professor at the University of Oxford, argues (1995, p. 40) that science became a tool of justification for biological explanations of the division of social roles by sex. She notes that in the 15th century, there was a resurgence of the Aristotelian view of women as lesser males, but this was dropped in favour of a view that had a better fit with the medical and anatomical knowledge of the time, which highlighted the differences between males and females. The result was a promulgation of anatomically or biologically based arguments intended to demonstrate the inferiority of women, focusing on brain size or other physical features, and on women's reproductive role. In this way, the development of science itself provided a new set of justifications for structuring society along sexual lines. Whether we see the rise of complement theories as a patriarchal reaction to a change in the intellectual climate that, for a moment, had opened up opportunities for women, or as arising in tandem with sociopolitical events that invited further division between men's and women's social roles, or as a result of the biological and medical knowledge of the time, the outcome is the same: a new way of justifying the divided social roles of men and women was born, along with a new tool, a developing science, to support it.

Furthermore, new sciences and pseudosciences, such as craniology, were developed and employed to support the general view that the differences between genders and races had a natural basis. In the 19th century, for example, craniometrists used special tools to measure skull size. The assumption was that skull size was related to brain size, which in turn was related to intelligence. Brains were also weighed and measured during post-mortems. The discipline was revived by the Nazis, who sought to scientifically justify both white supremacy and male supremacy. Criminals and those of non-white races were similarly believed to have inferior intelligence, and the data also seemed to point to women's mental inferiority, their skulls and brains being generally smaller than those of males. When new data showed that mass (weight by size) pointed to Asian people as being more intelligent than white people, struggles to demonstrate the superiority of white males became more and more confused and difficult. The fact that women generally have smaller bodies than men, and that skull and brain size are related to body size rather than to intelligence, was rationalized away (Gould 1981, p. 104). The argument then took on other, less obviously falsified forms, but it has not abated.

THE DISTINCTION BETWEEN NATURE AND NURTURE

Plato's claim that women with abilities ought to be afforded the same education as men points the way to a distinction used today, and still under debate, about a person's *nature* versus how a person is *nurtured*: that is, the distinction between our innate abilities and how we are groomed, moulded, and developed by the family and the society to which we belong.

Common to the views in which women were taken to be less intelligent or simply less capable than men is the assumption that what one sees is a product of nature, not of society. Thus, if it is observed in society that many women are more swayed by emotion, as, for instance, David Hume contended, it is because of their nature. If they lack the capacity to reason at a higher level, that is because they possess female brains. If they are just less intelligent, that is because their brains are smaller.

Plato did not make these assumptions about the nature of women. He found women in Athenian society to have "small intellects," but he did not suggest that this was natural (Averroes 1974, p. 496). Instead, he reasoned from the observed fact that women were talented in medicine and other disciplines to a position that women shared all the same abilities as men. In his view, while there were no Athenian female rulers in his time, there could have been. Instead of looking at differences in society, Plato focused on the ways in which women were similar to men in order to argue for the equality of their innate intellectual abilities. In contrast, thinkers such as Aristotle, Kant, and Hume focused on their observations of differences between men and women and used these to argue for women's innate inferiority with regard to certain intellectual abilities.

The point is that one can focus on different facts in society to support different arguments. It is a fact, for example, that there were no women rulers in Athens, but it is also a fact that there were women healers. The issue is whether women's actual social roles at the time were a result of the way society was structured or a reflection of their natural abilities. Plato argues that some women have the same abilities as men, but require the opportunity to nurture it. Others, such as Aristotle or Kant, would argue that there are few women philosophers or scientists because they simply lack the natural mental ability. In their view, the education of girls and women would be a waste. When confronted with arguments of this type, it is helpful to make the distinction between nature and nurture, and to point out that women have not had the same opportunities as men for thousands of years, and that can explain their lack of past achievements. It was precisely this approach that Mary Wollstonecraft adopted when, responding to Rousseau in her famous book *A Vindication of the Rights of Woman* (1792), she inaugurated modern feminism.

MARY WOLLSTONECRAFT

Like Mary Astell, Mary Wollstonecraft (1759–1797) accepted the claim that women could possess vices, but similarly pointed to upbringing and social restraints as the causes (as quoted in Waters 2000, pp. 97 and 105). In response to Rousseau, Wollstonecraft employed an argument similar to Plato's—in fact, she duplicated elements of the debate between Plato and Aristotle from over two thousand years earlier.

Like Plato, Wollstonecraft looked at the distribution of "virtues" between men and women in terms of degree rather than in terms of a supposed difference in kind. At the time Wollstonecraft was writing, "virtue" was a broad idea including intelligence and wisdom, and was not just a moral quality. On this question Wollstonecraft declares (1975, p. 139),

> I here throw down my gauntlet, and deny the existence of sexual virtues, not excepting modesty. For man and woman, truth, if I understand the meaning of the word, must be the same; yet the fanciful female character, so prettily drawn by poets and novelists, demanding the sacrifice of truth and sincerity, virtue becomes a relative idea, having no other foundation than utility, and of that utility men pretend arbitrarily to judge, shaping it to their own convenience.

Plato had argued that women and men, if they were virtuous, possessed the same kind of virtue: in other words, virtue did not change in nature when applied to a different gender, it just changed in degree. If a woman has less virtue, it does not mean that she possesses a different sort of virtue, less desirable, or, as Aristotle seemed to suggest, somehow a complementary to the virtues of men.

Wollstonecraft further argued that just because women have been treated as inferior, it does not follow that they are inferior. In other words, the existence of particular social relations is not an indication of their "naturalness" or unchangeability. Moreover, she asserts (1975, p. 155), "If woman be allowed [that is, admitted] to have an immortal soul, she must have, as the employment of life, an understanding to improve." Liberty, she argues, will make women more virtuous by creating the social climate in which their virtues may be nurtured through education and opportunity. In fact, as Kristin Waters (2000, p. 67) explains,

> throughout her work, she cleverly provides two types of arguments, one that appeals to the dominant ideology and one that appeals to the subversive one. This should be read not as inconsistency, but as rhetorical strategy. One kind of argument is that women who are educated to use their reason will be better wives and mothers. . . . The more radical argument holds that there is inherent and not just instrumental value in women developing virtue and reason. Women can, in a real sense, be equal to men.

JOHN STUART MILL AND WILLIAM JAMES

John Stuart Mill (1806–1873) also drew a distinction between women's natures and the way they are nurtured. Mill worked closely with Harriet Taylor, who eventually became his wife, and Helen Taylor, Harriet's daughter by her first marriage, and he was deeply influenced by both women (see Mill, Mill, and

Taylor 1994). In his essay *The Subjection of Women* (1869) Mill wrote (2001, pp. 14–15),

> The masters of women wanted more than simple obedience, and they turned the whole force of education to effect their purpose. All women are brought up from the very earliest years in the belief that their ideal character is the very opposite to that of men; not self-will, and government by self-control, but submission, and yielding to the control of others. All the moralities tell them that it is the duty of women, and all the current sentimentalities that it is their nature, to live for others; to make complete abnegation of themselves, and to have no life but in their affections.

While Mary Wollstonecraft had argued that it was women's lack of education that rendered them inferior, Mill pointed to the important role that education plays in shaping girls into submissive women, with characters that are formed to be the opposite of men's. Mill pointed out that a very important part of the socialization of girls and women includes fostering the general belief that such differences are natural. Women, he argued, are socialized to participate in their own subjugation: they are not just obedient but actually come to believe that it is not only their duty to live and do for others, but their true nature.

We noted earlier how rational arguments alone do not necessarily lead to social change: the first-wave feminism of the 19th century was tied to social movements such as the abolitionism in the United States and the labour unions in Britain and Russia. Mill also tied gender, race, and class together in his works. Just as Athens had been a democracy only for certain men, the ideas of equality emerging in the 17th, 18th, and 19th centuries were also not automatically extended to women. Thinkers like Mill, however, extended the scope of Descartes' claim that everyone possesses reason, and of Locke's claim that there is an essential equality between people. Mill argued (1994, p. 324) that we do not yet know women's nature because society has so constrained them: "I deny that any one knows, or can know, the nature of the two sexes, as long as they have only been seen in their present relation to one another." Like Mary Astell and Mary Wollstonecraft, he believed that "what is now called the nature of women is an eminently artificial thing—the result of forced repression in some directions, unnatural stimulation in others."

The American philosopher William James (1842–1910) charged Mill with holding a contradictory position (as quoted in Bell 1983, p. 355):

> In fact there runs through the whole book [*The Subjection of Women*] a sort of quibble on the expression "*nature* of women." The mainstay of his thesis is that there is nothing fixed in character, but that it may, through education of a sufficient number of generations, be produced of any quality to meet the demand; yet nevertheless he keeps speaking of woman's present condition as a distorted "unnatural" one. "Undesirable" is the only word he can consistently use.

For James, it made no sense to claim that people's positions in society are unnatural, there being no nature to accord with or go against. Yet one can still consistently argue, with Mill, that a woman's nature has not yet been discovered because of the social constraints women have been put under without being committed to claiming that they have no nature. In fact, contrary to James's interpretation, Mill does not seem to be claiming that people have no natures at all. He seems rather to believe that equal treatment and education will allow each person's real nature and talents to prosper. He writes (Mill 2001, p.17),

> It is not that all processes are supposed to be equally good, or all persons to be equally qualified for everything; but

that freedom of individual choice is now known to be the only thing which procures the adoption of the best processes, and throws each operation into the hands of those best qualified for it.

James was wrong to take Mill to mean that people have no natures. After all, if Mill did believe that, then he would think that, through education, every person could be "equally qualified for everything." Mill's argument is that while there may be many people who are equally qualified, there will also be cases where, in spite of equal education, some people will excel in areas where others will not. What Mill wanted was equal opportunity for everyone. In his essay he appeals to a higher principle of self-government for all individuals, arguing that society ought not to ascribe "natures" to people on the basis of their gender, race, or other accidents of birth. Mill is not saying that society ought to be based on people's natures, predetermined by sex. He appeals rather to the higher moral principle that all people ought to have a right to self-government, and this includes the opportunity to foster whatever innate talents an individual may have, through education, training, and personal development. Individuals should be allowed to determine their own positions throughout their lives rather than having society dictate to them. Mill sees that society and education can be the cause of women's subjugation and apparent inferiority, but that they could also be great equalizers, if everyone is given the same opportunities.

In any case, William James himself argued that women were essentially formed by the age of twenty and functioned through intuition. At the same age men were, in his view, inferior to women, but only because their brains were quite different and developed later. The masculine brain, he wrote (as quoted in Bell 1983, p. 354) "deals with new and complex matter directly by means of these [principles and heads of classification], in a manner in which the feminine method of direct intuition, admirably and rapidly as it performs within its limits, can vainly hope to cope with." What James means is that women simply receive their sensory information from the world, directly intuited. Men do so as well, but, according to James, they also possess a higher faculty that organises such data under "principles and heads of classification," James's term for what is usually associated with mathematics and science, especially in disciplines such as theoretical physics. Thus, while Aristotle relegated women to the world of the physical and emotional, James relegated them to a world of admirable but limited intuition. While Aristotle explained women's differences biologically, in terms of reproduction, James did it psychologically, through their purportedly different mental abilities.

GENDER DIFFERENCES: REAL, IMAGINED OR CREATED?

All the arguments or assertions seen so far involve the claim that women are different from men. The difference, it is claimed, is grounded in "nature," whether interpreted in terms of biology (Aristotle), inborn mental abilities (Hume, Kant, and James), or brain size (craniologists). Such natural differences were used to define what women could and could not do, and what they should or should not do. Yet, before determining whether natural differences ought to determine the structure of society, the prior question is whether the differences that these thinkers claimed existed between men and women are real, imagined or created.

Let us assume that some real differences exist. Recall how Plato's argument that some women share the capacity to rule with some men justifies his further claim that those women have a right to higher education. What this argument raises is the role of society in shaping capacities. In what follows, we look at responses to the arguments and claims that women are inferior, using the concept of created differences as opposed to real ones. This view brings into prominence the role of society, as opposed to the role of the isolated individual.

Plato argues that women, like men, ought to be afforded the chance, through education, to develop their natures in order to create the best society. Mary Astell and Mary Wollstonecraft argue that women, like men, possess the faculty of reason and ought to be afforded the chance to cultivate it and become better people. If women seem inferior, they argue, it is because they have not had the opportunity to cultivate their minds and talents. Mill is in the same camp, but he focuses on the role that society has played in cultivating not only what has been wrongly taken to be women's inferior "nature" but also how society and education fostered the complicity of women in their own subjugation.

Plato's aim was to build the best society around people's natural talents. He also argued that those who were naturally greater should govern those who were naturally lesser. In contrast, Mill was concerned with individual freedom and saw the best society as one in which each individual is self-governing and free, which would not have been the case in Plato's non-democratic ideal society. Astell and Wollstonecraft, like Averroes in the Middle Ages, added a moral component to their arguments: denying women education will lead to vice and an unworthy focus on what Astell calls "Trifles and Gaities." They argue that society will be improved by morally improving individuals, and part of this improvement includes access to higher education.

Astell, Wollstonecraft, and Mill thus bring to light the nurture side of the nature–nurture debate, in that they see society and education as essential contributors to the development of women's potential. Each of them also disagrees with the view that women are by nature inferior, and their arguments supporting access to education for women and their ideas of what constitutes the best sort of society rest on different premises.

It was suggested earlier that it was not for lack of reasoned argument, or of observed facts, that women were excluded from education and the public sphere. Plato offered a good reason for allowing women to be educated like men: the differences between men and women are not relevant to the work they might perform in society. Aspasia of Miletus is a ready example of a woman who, given access to education, could develop talents equal to, and even exceeding, those of some of her male counterparts. Following Descartes' split of the mind from the body, it was argued by Locke and others that the mind has no sex and that differences are located in the body. Logically, extending Locke's arguments would lead to the conclusion that equality ought to be extended to women, not just to property-owning men. Thinkers like Averroes, Mill, Wollstonecraft, Poullain, and Astell all pointed out the reasons why women *seemed* inferior in nature, and the ways in which society shaped them by favouring boys over girls, and consequently men over women.

There is no doubt that women have made progress. However, history shows us that women's advancement is not guaranteed and can be affected by a number of factors, political, economic, or ideological. The arguments seen today in, for example, the field of evolutionary biology studies are really not that different from Aristotle's arguments. Yet we can now see the power of arguments such as those of Wollstonecraft, Astell, and Mill, who all pointed to the crucial role played by society, and we now know the limits of a response such as Plato's—in spite of its insightfulness at the time—since it based the concept of equality on the fact of limited equality.

PETER SINGER: DO THE DIFFERENCES MATTER ANYWAY?

The contemporary Australian philosopher Peter Singer makes a distinction between the moral principle of equality and the factual equality that may exist between different races or between the sexes. In the cases we have examined from Plato onwards, with the exception of Mill, the moral principle of equality was tied to whatever factual equality or inequality was prevalent at the time. The idea was that if women and men were equal or unequal in nature, then society ought to mirror that. Of course, on this view, differences between the sexes (or races) could justify a non-egalitarian society. Even if we take Plato's view that the differences are negligible, or are spread equally between the sexes—that is, that not all men and not all women possess virtues in equal degrees—there is no guarantee that there may not be important differences between genders or races overall. In this regard, Singer (1999, p. 570) writes,

> So far as actual abilities are concerned, there do seem to be certain measurable differences between both races and sexes. These differences do not, of course, appear in each case, but only when averages are taken. More important still, we do not yet know how much of these differences are really due to the different genetic endowments of the various races and sexes, and how much is due to environmental differences that are the result of past and continuing discrimination. Perhaps all of the important differences will eventually prove to be environmental rather than genetic. Anyone opposed to racism and sexism will certainly hope that this will be so, for it will make the task of ending discrimination a lot easier; nevertheless it would be dangerous to rest the case against racism and sexism on the belief that all significant differences are environmental in origin. The opponent of, say, racism who takes this line will be unable to avoid conceding that, if differences in ability did after all prove to have some genetic connection with race, racism would in some way be defensible.

Singer's point is nicely summed up by Keith Burgess-Jackson (2002, p. 27): "Morally speaking, it doesn't *matter* whether men and women are the same or different, environmentally *or* genetically." Equality is a moral principle that need not be based on, or tied to, factual equality. As a moral principle, it is related to how we see and value human life, not on capacity or ability. By extracting our arguments from the nature–nurture debates and following instead a view such as Singer's, we have all the tools we need to ensure the continuation of the task of creating a society in which all persons are equal. Women have a long history to overcome, but also a rich one that can provide confidence and strength to continue to persevere and overcome obstacles. (A framework organizing arguments in terms of women's nature, the logic of their social position, their child-bearing capacity, the structure of society, and philosophers' views that can be used to support the position may be found in Appendix 1.)

PART II

Scientific Education of Women from the 17th Century to the 19th Century

———∞∞∞———

CHAPTER 4

Women Who Participated in Science in Early Modern Europe

Before we look at the struggle for equality and education, especially in science and mathematics, in early modern Europe, it is useful to examine the culture of science as it developed in the Middle Ages and during the "Scientific Revolution," leading to modern science as we know it today. This culture can be defined by several aspects of the scientific enterprise: who does science, the attributes and characteristics of its workers, the attributes that are valued and respected, and those that are not. The language and metaphors describing scientific approaches and activities are also of interest. (In more recent times, the choice of research questions that are likely to be funded, published, and rewarded has also become a key driver of modern science, as will be discussed in a later chapter.)

ORIGINS AND DEVELOPMENT OF SCIENTIFIC CULTURE

In most epochs, few women have been involved in science, philosophy, or mathematics, although there have also been times when the opportunity for women to participate more fully in these intellectual endeavours was greater than in other periods. Margaret Alic (1986, pp. 20–50) writes, for example, about women who were involved in philosophy, mathematics, and invention in various cities in ancient Greece, Rome, and Egypt. However, from the beginning of recorded history, science, mathematics, and philosophy have been domains of knowledge mainly constructed by men, for men, with the deliberate exclusion of women. It is not surprising that the culture of science has been, and remains to this day, masculine.

David F. Noble (1992, p. 3) has traced the historical origins of modern scientific culture by studying the evolution of the institutions associated with science: the professional societies, the academies, and the universities. He argues that this "world without women" did not simply emerge, but was constructed. An important factor was the rise, starting in the 2nd century CE, of clerical asceticism, which upheld traditional patriarchal patterns of female subordination and the exclusion of women from education. Much of the intellectual work, including science and mathematics, was performed by clerics and monks, except during a brief period in the 7th and 8th centuries, when double monasteries existed, with monks and nuns working side by side, in France, England, Ireland, and, later, Germany. These monasteries were considered centres of education and learning for women and men alike, and female intellectualism flourished during this period. The custom was for these double monasteries to be run by women (Noble 1992, pp. 30–33). Unfortunately, their decline and demise came with the Viking invasions and then the succession of monastic reforms brought about by Louis the Pious, son of Charlemagne, and St. Benedict. Only five of the fifty-three monasteries for women in France survived, virtually all of them being for cloistered women. Another negative factor was Charlemagne's insistence that all scholarly work, such as the writing of textbooks, sermons, biblical commentaries, theological treatises, and interpretations of canon and civil law, be done only by men. Double monasteries reappeared briefly during the religious revival of the 12th century, in a variety of forms, but early in the 13th century monasteries for women ceased to exist (Noble 1992, pp. 37–39).

The first western universities were founded in Paris, Oxford, and Bologna at the end of the 12th century, creating another "world without women." The new institutions, arising from ecclesiastical schools, were exclusively for males and entrenched the masculine culture in science and learning that has lasted for hundreds of years since.

In the Middle Ages, women became involved in alchemy, herbal medicine, and midwifery, and a few participated in what we would now call science in the fullest sense. Take the example of Hildegard of Bingen (1099–1179), who mentioned heliocentricity nearly four hundred years before Copernicus, speculated about universal gravitation five hundred years before Newton, composed numerous pieces of music, and wrote on medicine and natural history. Hildegard was relatively prominent in the hierarchy of the church and was canonized after her death, although more because of her religious visions than her scientific work, so her name and her achievements have survived the centuries.

Many other women who made, or could have made, contributions to science were not as fortunate as Hildegard. The European witch hunt, which resulted in the imprisonment and death of many women (Noble 1992, p. 208), began with the Catholic demonization of witchcraft through a papal bull issued by Innocent VIII in 1484, and was particularly intense in the 16th and 17th centuries, in France, Italy, England, Germany, and beyond. The feminization of witchcraft provided the momentum to effectively remove women from the roles they had played in medicine, alchemy, and other sciences, as Noble (1992, p. 210) explains: "Physicians played a prominent role in the witchcraft persecutions and the attack on the alchemists, perhaps in order to eliminate competition from lay healers and midwives." A few male alchemists suffered the same fate as women: in France, the ratio was said to be fifty females to one male. Noble (1992, p. 208) provides examples of what was done: "Nine hundred women were killed in a single year in Würtzburg, and over a thousand in Como . . . In Trier, two villages were left with only one woman each; in Toulouse (a Cathar stronghold) four hundred women were murdered in one day." As Carolyn Merchant (1980, pp. 138 and 168) points out, religious, social, and sexual attitudes toward women and their roles in society played a significant part in determining who became the victims.

THE IMPACT OF THE "SCIENTIFIC REVOLUTION"

It was against this background that, it is often said, the "scientific revolution" took place. The problem is not only that it was not called that at the time but that historians still debate whether what occurred amounted to a revolution at all. Science was referred to as natural philosophy until the 19th century; the word "scientist" was not coined until 1833. As Patricia Fara (2004, p. 23) points out,

> The Scientific Revolution, seen as so important for decades, is being ironed out of existence. Historians now emphasise continuity rather than change. They are challenging the story that science erupted suddenly in early modern Europe and are replacing it with different versions of the past. Many experts see a gradual transition around the turn of the 18th and 19th centuries, when modern scientific disciplines were being created, the financial rewards of research were being recognized, and research was starting to move out of people's private homes into laboratories, museums, and hospitals.

The debate over the term "scientific revolution," which involved a number of historians including the late Richard

Westfall and the late Betty Jo Teeter Dobbs, has been summarized by Margaret Osler (2000, p. 5) as follows:

> Dobbs challenges a traditional assumption about the heroes of the Scientific Revolution, namely, "that their thought patterns were fundamentally just like ours." It is only because they make this assumption that historians have found it difficult to explain Kepler's Pythagoreanism or Newton's devotion to alchemy. By making a different assumption, namely that people have not always viewed the world in the same way that we do, Dobbs is able to argue that we can make sense of their diverse interests and preoccupations. This is the crux of Westfall's disagreement with Dobbs. He assumes that thinkers in the past are similar to us, and that what is important for the historian is that aspect of a thinker's work that has survived until the present or that has led to our present way of looking at things.

In *Rethinking the Scientific Revolution*, the book Osler edited, Margaret Jacob argues that the disagreement between Dodds and Westfall was badly posed, since it rested on tacit assumptions about who and what made the scientific revolution. Jacob's conclusion is that there *was* a scientific revolution, but not the one that both Dobbs and Westfall tacitly assumed was at stake. She proposes that the revolution was carried out in the 18th century, when, as Osler puts it in her introduction to the book (2000, p. 5), "natural philosophers selectively took up Newton's physics and mathematics while ignoring his alchemical and theological views."

The details of the historians' debate lie beyond the scope of this book, but of relevance here is the fact that historians of science have, for the most part, ignored women's contributions to the achievements of the early modern era, as well as the well-founded argument that the serious steps taken at the time to formalize scientific activity excluded women from these new circles of power, and had a serious and deleterious impact on women's future involvement in science.

MEN'S CONTRIBUTIONS

The "age of discoveries," starting in the late 15th century, had featured the search for spices and other exotic products, developments in mining and metallurgy, a general expansion of commerce and trade, and, above all, the spread of printed books, making reading materials available to much larger numbers of people than ever before. Against this background, historians of science have tended to focus on the achievements of a few male figures as central to the progress that was then made in the 16th and 17th centuries: the replacement of the Aristotelian–Ptolemaic theory that Earth is the centre of the universe by the heliocentric theory put forward by Copernicus (1473–1543); the mathematical explanations of planetary motions attributed to Tycho Brahe (1546–1601) and Johannes Kepler (1571–1630); the development of the concept of experimental science by Sir Francis Bacon (1561–1626), Galileo Galilei (1564–1642), René Descartes (1596–1650), and Blaise Pascal (1623–1662); and the new mathematical theory proposed by Descartes and applied to physics by Sir Isaac Newton (1643–1727).

Parallel developments occurred in medicine, with the writing by Andreas Vesalius (1514–1564) of the first complete textbook on human anatomy, *De Humanis Corporis Fabrica*, and the construction by William Harvey (1578–1657) of an accurate theory of how the heart and the circulatory system operate, published in his book *On the Motion of the Heart and Blood in Animals* (1628). The invention of instruments was another indication of intense scientific activity: the telescope, which stimulated many discoveries in astronomy; the microscope, which came to be extensively used in biology and botany; the first precision clock, which allowed the measurement of time with great accuracy; as well as the thermometer and the barometer. Robert Boyle (1627–1691), with the assistance of Robert Hooke (1635–1703), devised the vacuum chamber or air-pump in which he was able to study the nature of air. Boyle's law, which he subsequently formulated, describes mechanical phenomena in the study of fluids, and the relationship between the volume and pressure of gases.

R. S. Westfall (1971, pp. 113–14) is by no means the only historian to have surveyed these and other developments, and concluded that no previous century had made such important contributions. In this period, nature was mathematized and mechanized. A discontinuity opened up between the Aristotelian and Newtonian views of the world, and natural philosophy became increasingly removed from the hermetic philosophy that had been favoured by Paracelsus and his followers in the previous century. Experimentation became an important aspect of supporting scientific theories and laws.

The second half of the 17th century witnessed the emergence of several new academies dedicated to scientific study, creating a means of communication for science outside the traditional setting of the universities (McLellan 1985, p. xix). The Royal Society was created in London in 1662, the Académie Royale des Sciences in Paris in 1666, and the Societas Regia Scientiarum in Berlin in 1700. By the middle of the 18th century these and other academies, numbering seventy or more across Europe and the Americas, had become powerful centres for the accumulation of specialized scientific knowledge.

The concept of the scientific journal, in which scientists publish and review each other's ideas and findings, began with these learned societies, as James E. McLellan (1985, p. xix) explains:

> scientific societies were centralized clearing houses for communicating science . . . the network and system of institutions established *de facto* by interaction among the

scientific societies provided an entirely new institutional means of producing, channelling, and facilitating the exchange of scientific information.

The Académie des Sciences, the Royal Society, the Berlin Academy, and many of the other learned societies had one thing in common: except at Bologna and certain other Italian city-states, they were usually reserved exclusively for men. Thus the crucial hub for scientific activity and the presentation of new work excluded women. In fact, the monastic culture was transposed into the scientific societies when they emerged. Robert Boyle, one of the first members of the Royal Society, even took a vow of celibacy at the age of twenty-one and devoted himself to God and to science. The revivalist spirit of the Reformation rekindled asceticism, and the Royal Society distanced itself from anything feminine, as this could be associated with witchcraft and alchemy. In 1664 the first secretary of the Royal Society, Henry Oldenburg (1619–1677), announced that its purpose was "to raise a Masculine Philosophy . . . whereby the Mind of Man may be ennobled with the Knowledge of Solid Truths" (quoted in Rose, 1994, p. 17; see also Easlea 1981). As the Restoration in 1660 re-established the monarchy and the Church of England, so the Royal Society could be said to have restored the clerical and the academic, and hence male monopoly over natural philosophy (Noble 1992, p. 230).

The universities did not then offer an alternative route to scientific education and achievement. Few of them were involved in teaching science or performing scientific research, though Newton, for example, was appointed to the Lucasian Chair in Cambridge in 1669 and was able to pursue his work on optics there. These centres of higher learning also excluded women until late in the 19th century, again with a few exceptions in Italy (see Chapter 6).

However, in spite of their near-total exclusion from formal study and research in the academies and universities, some women also contributed to the making of important scientific discoveries and achievements. Now that more information is coming to light about women's participation in these activities, it is possible to uncover the work not only of some of the men who have undoubtedly been forgotten by mainstream history, but also of the many women who, provided with a good education, tools, and the support of a male relative, such as a husband, brother, or father, were also able to undertake serious science.

WOMEN'S CONTRIBUTIONS AS SCIENTISTS

Scientific discovery, even in the early modern era, was rarely the achievement of a single person. Behind the person who was given the credit, both then and now, were many artisans and assistants, women and men, who contributed to the work. Women scientists were predominantly from the nobility or the bourgeoisie, working with brothers, fathers, or husbands. These men provided access to equipment, such as telescopes, and the women provided assistance in the work itself, not only by collecting data and making observations but by contributing to the development of mathematics and astronomy.

Many of the women about whom information has become widely available only recently worked in astronomy, and several made important contributions to that science. It is perhaps surprising to learn that in the 17th century an estimated 14 percent of astronomers in Germany were women (Schiebinger 1989, p. 66). Although their lives and achievements still remain relatively invisible, the development of women's studies programmes and of the World Wide Web have helped to uncover and spread information about them. This section presents a few examples of such women (drawn from Schiebinger 1989, pp. 47–99).

The Danish astronomer Sophia Brahe (1556–1643) became passionate about the subject at the age of ten, while she was assisting her brother, Tycho Brahe, in his observatory, Uraniborg, on the island of Hven. Their parents provided her with a private tutor in mathematics, music, alchemy, and medicine, but her work with Tycho was interrupted by a forced marriage in 1576, when she was twenty, to a man aged thirty-three, and the birth of a son in 1580. After her husband died, in 1588, she rejoined Tycho and together, over many years, they recorded the positions of planets and stars, and made computations to predict eclipses and comets. Sophia was also a horticulturist, a healer, and a historian, and became something of a legend in her own lifetime. Although she collected a great deal of data and made important observations, often on her own, it is her brother who is still generally credited with their work, is the subject of numerous books, and has a planetarium in Denmark named after him.

In Germany, Maria Eimmart (1676–1707), working in her father's observatory, prepared 250 drawings of the phases of the moon in a continuous series that laid the groundwork for a new lunar map. She also made two drawings of the total lunar eclipse of 1706 and may have written the whole or parts of a paper about the sun that was published under her father's name. Maria married Johann Heinrich Muller in 1706, and he inherited her father's observatory. She continued her work in astronomy after her marriage, but her career was cut short when she died in childbirth in 1707 (Schiebinger 1989, p. 81).

Marie Cunitz (1610–1664), born in Silesia, was the daughter of a doctor who taught her six languages (Hebrew, Greek, Latin, Polish, Italian, and French) in addition to their native German. She also studied music, mathematics, and medicine. After her marriage to a local doctor, she involved herself in history, poetry, painting, and music, as well as her principal interest, astronomy. Her translations and modifications of Kepler's laws, published in *Urania propitia tabulae astronomicae mire faciles* (1650), made Kepler's work more accessible to scholars, and became the only ones available for many years after her death (Schiebinger 1989,

p. 80). Although Maria Cunitz's work is highly regarded by some, she was severely criticized many years after her death for neglecting her womanly duties because of her work in astronomy.

Maria Winkelmann (1670–1720), born near Leipzig, assisted one astronomer, Christoph Arnold, before marrying another, Gottfried Kirch, who was some thirty years older than herself. She helped her husband with calculations, observations, and the making of calendars, and in 1700 moved with him to Berlin. In April 1702 she discovered a previously unknown comet, which should have secured her a position in the astronomical community, especially as her husband's own position was partly based on his discovery of a comet in 1680 (Schiebinger 1989, p. 85). Unfortunately, the report of her finding and a subsequent publication regarding this discovery bore only her husband's name. In 1710, however, Kirch himself recorded in his notes that his wife had discovered the comet while he was sleeping. As a consequence of this omission, historians studying original documents have continued to attribute the entire work to her husband. Another of Maria Winkelmann's contributions was the prediction of the appearance of a comet in 1712, achieved through observations and accurate calculation. After her husband's death she continued as assistant to her son Chriestfried, again at the Berlin Academy, into which she was never formally admitted. Instead, she was forced out on October 21, 1717, because she had not heeded warnings to retire to the background when visitors came and to leave the talking to the men. However, the Academy did not want her to give up her duties as a mother and expressed the hope that Maria "could find a house nearby so that [her son] Herr Kirch could continue to eat at her table" (Schiebinger 1989, p. 97). Maria's daughters Christine and Margarheta were also competent astronomers, but their role was restricted to assisting their brother. Londa Schiebinger reports that, unlike their mother, the daughters "did not ask for official positions. Nor did they exude the fire of their mother, badgering the academy for housing or greeting foreign [male] visitors" (Schiebinger 1989, p. 99).

Finally, Margaret Cavendish, Duchess of Newcastle-upon-Tyne (whom we have already met in Chapter 2), continued her intellectual development after her marriage to William Cavendish, who had some knowledge of mathematics and science. The publication of her numerous books was financed by her husband, who added a laudatory verse to each of them: they include *The Philosophical and Physical Opinions* (1655), *Natures Pictures Drawn by Fancies Pencil to the Life* (1656), *Observations upon Experimental Philosophy* (1666), and *Grounds of Natural Philosophy* (1668). Margaret Cavendish participated in discussions with well-known philosophers on subjects such as matter and motion, the existence of the vacuum, the nature of magnetism, life and generation, colour and fire, perception and knowledge (Schiebinger 1989, p. 47). She was also the first woman to visit the Royal Society, an incident that created a huge controversy, since women were barred from membership. Interestingly, Cavendish does not mention the visit in her autobiography, as the experience had been less than positive: it is said that she was demeaned and ridiculed while she was there. In her writings, too, her approach to nature was quite opposite to that taken by the famous male philosophers of the period, and, as Londa Schiebinger writes, her "rejection of a sharp distinction between animate and inanimate nature led her to reject the Baconian approach that man through science should become master and possessor of nature." She also made bold criticisms of Hobbes, Descartes, and others. She was severely criticized by many, and her response to these attacks was to apologize for her lack of a formal education.

WOMEN'S CONTRIBUTIONS AS PATRONS AND CORRESPONDENTS

Another role played by women in the early modern era was as patrons and correspondents of men involved in science, although this was mainly the case for women from royal or noble families.

For example, Elizabeth Von der Platz (1618–1680), a Protestant abbess of Herford, was the eldest daughter of Frederick V and Elizabeth Stuart. She conversed and corresponded with René Descartes for seven years, pointing out some of the weaknesses and contradictions in his writings and forcing him to explain his ideas better (Fara 2004, pp. 66–69). Elizabeth also corresponded with Leibniz, Margaret Cavendish, and Anna Maria von Shurmann, a lecturer at Leiden University in the Netherlands, among others. Her daughter Sophia, Electress of Hanover (1630–1714) also corresponded with Leibniz (Fara 2004, p. 63). Sophia's daughter, Sophie Charlotte, Electress of Brandenburg and later Queen of Prussia, directed her minister Jablonski to build an observatory in Berlin, and worked closely with Leibniz and others to carry forward the plans to create the Berlin Academy. Caroline of Ansbach (1683–1737), wife of King George II of Great Britain, was also a major patron of science and a student of Leibniz, and held discussions with Newton.

Christina (1626–1669), Queen of Sweden between 1644 and 1654, was a patron of the arts, was interested in philosophy, and was Descartes' second and last patron (Fara 2004, p. 57). Although she was not a scientist herself, her interest in science, and her avoidance of marriage due to the fact that it restricted the role of women, added to her reputation for independence of mind.

In Paris, upper-class women could meet renowned scientists, philosophers, and writers in the salons that were then in vogue, and on certain days scientific experiments and demonstrations were given. The major scientific topics of interest to women who attended these gatherings were botany, physics, alchemy (a precursor of chemistry), and mathematics. For example, the weekly salons conducted by Madeleine de Scudéry (1607–1701) formed the centre of the cultural movement that swept through France in the 1650s.

These sessions have been aptly referred to as "science for a polite society." Nevertheless, women were ridiculed for their participation in these intellectual activities. For example, in his play *Les précieuses*, Molière (1622–1673) suggests that women attending salons were just trying to find charming suitors (Sutton 1995, p. 104).

THE MASCULINIZATION OF SCIENCE

As we have seen in this chapter, the early modern era, whether we call it a "scientific revolution" or not, was a time of great change, with not only the emergence of new scientific theories that contradicted past beliefs, but also the formalization of science through the academies, excluding women, who had to be contented with having husbands present their work, if they would.

Sir Francis Bacon had a major influence on how the new science was to be done, as Linda Jean Shepherd (1993, pp. 19–20) has emphasized:

> Francis Bacon advocated using the new experimental philosophy to inaugurate the "truly masculine birth of time", to lead men to "Nature, with all her children, to bind her to your service, and make her your slave, to conquer and subdue her; to shake her to her foundations" . . . Bacon urged researchers to use his method to discover the "secrets still locked in Nature's bosom, . . . to storm and occupy her castles and strongholds, and extend the bounds of human empire."

Bacon's metaphors clearly express control of man over woman, and nature, too, was often portrayed as a woman. These images unequivocally portray a masculine philosophy of science, which may well not have appealed to many women. In the same era, Joseph Glanvill, the chief promoter of The Royal Society in its early days, maintained that "truth does not have a chance when the Affections wear the breeches and the Female rules," and he insisted that the ideal scientist was unemotional and detached, for his tools were logic and analysis (Shepherd 1993, pp. 19–20). Most men, and many women too, seriously believed that women had no role to play in science.

These intellectual developments did not, of course, take place in a vacuum. From time to time, social divisions arose from divisions within political leaderships or social revolutions. For example, in England James I sanctioned what conduct was appropriate for males and females in 1620. These sanctions on modes of conduct infiltrated society at all levels, through sermons and pamphlets. Later in the 17th century, the Civil Wars in the British Isles resulted in a further division between men's and women's roles, in spite of all the discussion and promises of social reform, including the principle of universal education formulated as a goal by many of those who rose against the monarchy. Early industrial capitalism also had its impact: as mechanization moved production from the home to the workshop, women lost their independent wage-earning opportunities and were increasingly reduced to being dependent housewives (Keller 1985, p. 63). Evelyn Fox Keller has argued that science in the early modern era worked in tandem with this increased polarization between the genders by ascribing to itself all the constructed masculine attributes and none of the feminine ones. Thus science as an endeavour became masculinized, excluding women as practitioners. Science and masculinity became associated with mind, reason, and objectivity, while females were ascribed nature, feeling, and subjectivity. The assumption was that women were *by nature* not suited to the practice of science.

CHAPTER 5

Education for Women in the 16th and 17th Centuries

Education in the early modern era was gendered, meaning that boys and girls were provided with vastly different educational opportunities. In her book *Better than Rubies: A History of Women's Education* (1978) Phyllis Stock discusses the general link between education and the societal role of women, and suggests (pp. 12–13) that education

> has a social, rather than an individual function. The aim of the educator is to produce an adult who will play a certain desired role in society . . . Naturally, in any time when men and women were viewed as totally different beings, the education provided them also differed. . . . Thus, the education of the two sexes reflects actual social and power relationships between men and women in the society.

In this regard, a distinction needs to be made between formal and informal education. Barbara Whitehead (1999, p. x) makes the point that using the traditional, formal definition of education, especially in observing past eras, often excludes women and girls:

> Just as traditional history excluded women by defining the historical subject matter from a man's perspective, the traditional definition of education would exclude early modern women from the history of education. Early modern education, defined primarily as formal training in schools and universities with an emphasis on the learning of Latin, would be out of reach for nearly all women of that period.

For a broader understanding of the education of women and girls, Whitehead recommends including the different types of learning that women typically received in past eras, observing how these were linked to their social roles or vocations. Whitehead also provides (1999, p. xii) an example of what each gender was expected to learn in early modern Europe:

> An educated nobleman who sought access to the upper ranks of the royal bureaucracy was considered educated when he had a humanist education, whereas his wife, who could never aspire to such a position, would have been considered educated if she could sew, dance, play music, and run a household.

Phyllis Stock (1978, pp. 13–14) divides women's education into six general categories, many of which represent informal types of learning:

(1) informal moral education, usually combined with housewifery, provided by female members of the family or others *in loco parentis*;
(2) formal moral education, usually based in some knowledge of reading and provided by nuns or other women in schools and convents;

(3) vocational education in a trade, provided by parents, others *in loco parentis*, or husbands;

(4) intellectual education aimed at character formation, usually provided at home by parents or male tutors;

(5) intellectual education for the fulfilment of a particular role in society (court lady, mistress of the estate, first educator of the children) and supplemented with training in the social graces, provided by tutors, convents, schools; and

(6) intellectual training for individual fulfilment, a career, or both, usually provided in formal educational settings.

Stock observes that when all social classes are considered, up to 90 percent of early modern women were illiterate. This is confirmed by David Cressy, who reports that illiteracy among women in London in 1600 was about 90 percent, but notes that this was a lower rate than anywhere else in England at the time and was comparable to the rates of illiteracy among labourers and husbandmen. By 1640, however, illiteracy among women in London appears to have declined to about 80 percent, and by the 1690s it had fallen to 52 percent, while remaining at much higher levels in rural areas. Cressy suggests that the reasons for the increase in literacy in the towns and cities may have included the benevolent patronage of servants and apprentices by their masters, and a tendency for the more literate to migrate to London in search of work (Cressy 1980, pp. 128–29).

It was the exclusion of women from most kinds of work that most greatly influenced the type of education they received and the level of literacy they were able to attain. As we have seen in previous chapters, women's roles, defined by men over many centuries, were generally limited to the private realm of the home, but, as Phyllis Stock (1978, p. 13) writes,

there were also times when the patriarchal system weakened, when women became socially, economically, or

politically independent; then women freed themselves from
the male concept of them and to some extent controlled
their own intellectual life and values.

It was easier, of course, to find this freedom in the upper
classes, yet even in the middle and lower classes this may have
been the case for those widows who were able to manage their
late husbands' businesses. For most women, in all classes, lack
of education went hand in hand with lack of opportunities to
work outside the home.

EDUCATION OF WOMEN
IN THE 16th CENTURY

In the 16th century, as in earlier times, almost all girls learned
at their mothers' knees all they needed to know to become
wives and mothers, and to run households. The skills they
were expected to acquire included sewing, cooking, basic
medicine, household accounting, and, in the lower classes,
agricultural techniques. (O'Day 1982, p. 183). Reading, to
the extent that it was taught to girls at all, was primarily
provided so that women could read the Scriptures. Even
relatively liberal Renaissance educationalists were confined
by the common conception of a woman's vocation. Philosophy,
poetry, Latin, Greek, and rhetoric were considered inappropriate studies for women, as they might foster self-expression,
seduce women away from simple Christian truths, and arouse
a desire to participate in the public realm. The daughters of Sir
Thomas More received rigorous training in the classics, and
so did Lady Jane Grey, the "nine-day Queen," but they were
very much exceptions to the rule.

The arguments of Richard Mulcaster (c. 1531–1611), the
first headmaster of Merchant Taylors' School in London, and
later High Master of St. Paul's School, give a better indication
of the prevailing wisdom. Mulcaster argued that exceptional
women were mere curiosities, like guinea pigs or other exotic

animals, and set no pattern for their sex (O'Day 1982, p. 185). He approved of girls receiving a certain limited degree of formal education, but emphasized, in his book *Positions* (1581) and other writings, that they should learn only what they needed to know in order to fulfil their religious duties and prepare for their vocations as housewives and mothers. He took it for granted that girls had brains "less charged" than those of boys, although, "like empty casks, they make greater noise" (Fraser 1994, p. 153). His recommendation that boys and girls be educated together in elementary schools up to the age of nine was regarded by many of his colleagues as dangerously radical.

For most women, then, reading materials were carefully regulated and mainly consisted of conduct books. One particularly influential example was *De Institutione Feminae Christianae* by Juan Luis Vives (1493–1540, also known by his Latin name Joannes Ludovicus Vives). This conduct book, published in 1523, was dedicated to his patron, Catherine of Aragon, the first wife of King Henry VIII of England, who had commissioned it for the benefit of her daughter, the Princess Mary (later Queen Mary I). Phyllis Stock (1978, p. 52) describes the book's main themes:

> Vives set the tone for his treatise by noting that innumerable things must be taught to men, who are active at home and abroad, while women could be taught easily, since their only concerns were honesty and chastity. . . . [A girl] should learn cooking and spinning, as well as her letters. Vives found no defects in their ability to learn. Although educated women were often morally suspect, and learning might be an aid to deceit rather than to virtue, good women, if they were "fenced in with holy counsels," would not be harmed by knowledge.

Vives agreed with Erasmus that women should not be teachers. His attitude to women's learning shows clearly when he writes,

"For it cannot lightly be a chaste mayde that is occupied with thynknge on armour," meaning that a young woman cannot be chaste (as, of course, she was then expected to be) if she is preoccupied with stories about knights.

Although conduct books were initially directed at royal and noble women, they gradually filtered down to other social classes. *De Institutione Feminae Christianae* alone was published in forty editions and several translations were made into other languages, including English (as *Instructions of a Christian Woman*). The conduct books aimed directly at middle-class audiences reiterated the end to which women's learning was to be directed: their roles as mothers and wives. Thus, *La Institutione di una Fanciulla Nate Nobilmente* by Gian Michele Bruto (1517–1592), translated by Thomas Salter as *The Mirrhor of Modestie* (1579), stresses the importance of the acquisition of traditional female virtues such as chastity, piety, and humility, and Bruto, like Vives, reserves his greatest opprobrium for chivalric fiction and plays. The Bible, the teachings of the Church Fathers, and narratives of virtuous women are judged to be much more appropriate materials for imparting male constructs of female virtue to women.

In England, as in other countries that became officially Protestant, the Reformation had a dramatically negative impact on the education of women. Convents disappeared, nuns were no longer permitted to teach, and, while some of the wealthier Catholic families began to send their daughters to study in convents abroad, this was a dangerous undertaking in a time of religious upheaval and constant warfare. In 1621, the English House of Commons resolved that all students must be brought back from abroad to be taught at home by Protestants.

In Catholic and Protestant countries alike, theology glorified marriage, but it also glorified the patriarchal role of men as heads of families (O'Day 1982, p. 180). The Homily on Marriage, issued in England in 1562 and ordered to be read in all the churches, emphasized duties of love, honour, and obedience of wife to husband, and similar documents were officially imposed in other countries, where, in most cases, failure to attend church was a punishable offence. In spite of this, some wives with dominant personalities stood firm and resisted their husbands, and in some households husbands discussed public issues with their wives before making decisions. Women generally had some say in the running of households, but no rights or role in public life (O'Day 1982, p. 183).

In close collaboration with the churches, the royal families of Europe also had an important impact on the education of girls and women, being expected to lead by example. The level of education received by female members of royal families varied: Mary, Queen of Scots (1542–1587), Mary I of England (1516–1558), and Elizabeth I of England (1533–1603) each received a broad classical education in Latin and Greek, but few other royal women did, and none of these three monarchs did anything to challenge social convention or promote the education of other women. James I, the estranged son of Mary, Queen of Scots—who declared, "To make women learned and foxes tame has the same effect: to make them cunning," and forbade his daughter Elizabeth to learn Latin (Fraser 1984, p. 136)—is a much more representative monarch of the time.

MARIE DE GOURNAY

Even in the 16ᵗʰ century we can find at least one extraordinary woman who lived by her pen, and even supported her mother and her siblings after her father's death. Marie de Gournay (1565–1645) became well known in her lifetime after she wrote the preface to the *Essais* of Michel de Montaigne (1533–1592), helping him to advocate a humanistic morality that contrasted with the religious intolerance of their times (Gournay 1988, p. 9).

Marie de Gournay understood when still relatively young that the only education available to women, beyond the basics taught in the home, was self-education. Accordingly, among other accomplishments, she taught herself Latin by comparing French translations with the original Latin texts. She went on to publish two treatises, *Grief des dames* (1595, expanded 1626) and *Égalité des hommes et des femmes* (1622), as well as an autobiography and essays on the education of children in France.

In *Grief des dames*, Marie de Gournay complains about the lack of property, freedom, and access to public office for women, and she defends the right of educated women to be heard and to be read equally with men of letters. Like Descartes, and against the trend of her day, she separates mind from body and argues that women are as capable as men. In *Égalité des hommes et des femmes*, she uses an approach similar to that of Christine de Pizan in her *Book of the City of Ladies* (discussed in the Preface): profiling great women of the past to demonstrate women's abilities to learn. She avoids taking extreme positions and stresses the term "equality," as opposed to claiming the superiority of one sex over the other. She also rejects the idea that a great woman would simply resemble a great man. She argues that it is not surprising that women are found to be incompetent, ignorant, and focused on the physical aspects of their lives, given how little education and encouragement most of them receive. As Maya Bijvoet writes, Marie de Gournay

> focuses on the theological questions and searches the philosophers for evidence in favour of Eve's equality, for ideas to counter the low regard in which women's intelligence was generally held. . . . She believes that, given the same opportunities, privileges, and education usually granted to men, women can equal men's accomplishments. The discrepancy in intelligence and achievements between men and women results from differences in education, circumstances, and attitudes, not from an inherent, predestined, intellectual inequality.

Marie de Gournay received great honour in Brussels and in Anvers during her lifetime. She also found a patron in Marguerite de Valois (1553–1615), the wife of King Henri IV of France and Navarre, who invited Marie to her royal salon and provided her with quarterly financial support. Marie de Gournay considered her own writings equivalent to those of men of her time, with the exception of Montaigne, whom she considered a "father" and the greatest writer of the period. Her scientific interests were focused on alchemy, a popular activity at that time (Gournay 1988, p. 33).

EDUCATION OF WOMEN
IN THE 17th CENTURY

In the 17th century, as in the 16th, most women themselves were not interested in formal education for their daughters, believing, with good reason, that reading and writing skills would not support them later in life. Instead, the emphasis, in the home as well in schools for girls, was on practical skills. Some schools excluded girls entirely, such as Harrow School under its statutes of 1589, while Banbury School allowed girls to receive a basic education, but only up to the age of nine (O'Day 1982, p. 185). Boys, on the other hand, could progress from elementary school to grammar school and to university, if their parents had the means and they had the ability to master reading, arithmetic, and writing, and then Latin and Greek. When girls did attend school, they were not permitted to study Latin and Greek, and reading continued to be limited to Bible study and conduct books. Overall, the quality of formal education for girls may even have declined from the late 16th century onwards, at least in England, "as the practice by which a few girls had attended the grammar schools, if not to an advanced age, ceased" (Fraser 1984, pp. 152–53). Those few women who did secure a formal education then faced a dilemma: as Antonia Fraser (1984, p. 133) succinctly puts it, "They were not valued if not educated, but also they were scorned if they were." Whether they were educated or not, women were still expected to remain modest and obey their husbands (Fraser 1984, p. 137), a view bolstered in the 17th century, as in previous centuries, by the combined forces of church and state.

One example of how religion was closely tied to the education of women can be found in the work of Mary Ward (1585–1645). Born in Yorkshire but educated in the Low Countries, Ward ran several Catholic schools abroad and had an undying obsession with teaching piety, morals, and liberal arts to girls. The people of the Low Countries generally lived much freer lives than in England, and women engaged with men in conversation and debate. Ward asserted, "No difference between men and women that women may not do great things!" (as quoted in Fraser 1984, p. 153). However, her search for truth was not for truth for women, or truth for men, but truth for God, and the purpose of her teaching was to make women docile and grateful for their lot, rather than questioning it. The arrival in England of the French Catholic princess Henrietta Maria (1609–1669), and her marriage to King Charles I, may have seemed to bring promise of educational opportunities for girls, as she was a follower of Mary Ward. Yet even educated women of the time, such as Lucy Hutchinson and Margaret Cavendish, did not think highly of the efforts of Henrietta Maria or other royal spouses to influence policy. Lucy Hutchinson wrote that it was "an unhappy kingdom where the hands which were made only for distaffs affected the management of sceptres," and praised Queen Elizabeth I for "acceding to her male counsellors" (Fraser 1984, p. 152).

One positive move for the education of girls was the rise in the number of boarding schools for the prosperous middle class. The first such school for "young gentlewomen" in England was opened in 1617. However, the emphasis in these institutions was on acquiring social graces rather than displaying erudition. A form of shorthand was taught to some of these girls, but this was still considered part of the traditional set of women's skills (Fraser 1984, p. 157). Other schools for young ladies included two large establishments founded before 1650 by a Mrs. Salmon in Hackney, then a village outside London. They taught some knowledge of French, housewifery, and accounting. Still other schools, such as the one established by Mrs. Perwich in Bohemia Place in Hackney, provided music lessons. For girls of the upper classes, it was a happy accident if they came from a large family and had brothers who were tutored or who could teach them, as they could learn alongside the boys.

SCIENTIFIC EDUCATION FOR WOMEN IN THE 17th CENTURY

Science—or, as it was then called, natural philosophy—was not highly valued in the 17th century and was for the most part left out of the curriculum in boys' schools and in institutions of higher learning, in spite of the remarkable scientific developments that were occurring during this period (as discussed in Chapter 4). As James E. McLellan (1985, p. xxii) observes,

> Regarding the 17th century, all things considered, and despite the profound intellectual reformulations of the Scientific Revolution, science remained a relatively small-scale enterprise in society. . . . Science had yet to demonstrate its authority as a means of knowing.

The study of the classics was still the main focus of formal education, and science was regarded as an undesirable area of study for "gentlemen." As Patricia Phillips (1990, pp. ix–x) writes,

> For centuries an education in the classics was considered by the universities and by society in general to be an

intrinsically masculine avocation, to constitute the only
worthwhile field of scholarship and, furthermore, to confer
the essential stamp of gentle birth and good breeding.
The pursuit of science, on the other hand, was tainted by
its too intimate involvement with mechanical and menial
matters—undesirable on both scholarly and social grounds.

Phillips further argues (1990, p. x) that society's low esteem
for scientific pursuits actually provided an opportunity for
women to become involved in them:

While many gentlemen, members of the privileged elite,
repudiated science as too contemptible and too trivial
to warrant their own attention, some thought these very
demerits made it unobjectionable as a study for their
ladies. The operations of a laboratory, after all, were not
dissimilar to those of the kitchen, and scrutinizing lower
forms of life through a microscope was more womanly
than vain attempts to master the complexities of Latin
and Greek.

Remembering the debate in the *Lady's Journal* about women's
inferiority, Pierre Antoine (Peter) Motteux (1663–1718)
thought that the interest in the classics by boys, and the lack
of it by girls, could be a good thing for the girls. Motteux
suggested that it was women's good fortune not to be educated
in the classics, and not, as often claimed, a negative side of
women's character. He concluded (as quoted in Phillips 1990,
p. 23), "With such creative temperaments, it was clear that they
made finer authors and thinkers than the supposedly better
educated and superior male." Phillips also suggests (1990,
p. 29) that being excluded from a classic education created
opportunities for women to become interested in science:

To women seeking an outlet for their intellectual
aspirations, the independence and originality of science

and scientists must have seemed attractive. In this field, female ignorance of the classics and the originality and unfettered imaginative powers with which women were said to be endowed were positive advantages.

Even where the social profile of science was rising, some women could still make the most of the opportunities available. Geoffrey Sutton (1995, p. 116) points out that in France,

> the study of nature in general had become more visible and more accessible during Colbert's ministry. The philosophical *matinée* was no longer the only regular source of scientific intercourse (where men and women could attend). In 1665, the *Journal des sçavans* appeared (with more or less regularity) every week and later every fortnight, for sale at the printer or at a growing number of distribution points. The French periodical covered many fields and was a good source of information for women.

As the century went on, proposals for educational reform tended increasingly to advocate the inclusion of science in the formal curriculum, and the extension of formal schooling to women. Many such proposals were influenced by, or were responses to, the writings of the Czech reformer Jan Amos Comenius (or Komenský, 1592–1670), who was invited to visit England by Samuel Hartlib, John Drury, and Robert Boyle, and also found followers and supporters in Germany. Comenius's philosophy of education is to be found in the three volumes of his work *Didactica magna* (1633–1638, *The Great Didactic*), in which he insists upon complete equality of the sexes, and argues that everything must be taught to everyone. Piaget (1993, p. 9) quotes Comenius (1896, pp. 219–20) as follows:

> Nor can any good reason be given why the weaker sex (to give a word of advice on this point in particular) should

be altogether excluded from the pursuit of knowledge (whether in Latin or in their mother-tongue). . . . They are endowed with equal sharpness of mind and capacity for knowledge (often with more than the opposite sex) and they are able to attain the highest positions, since they have often been called by God Himself to rule over nations, . . . [and] to the study of medicine and of other things which benefit the human race. . . . Why, therefore, should we admit them to the alphabet, and afterwards drive them away from books?

Comenius's views stand in stark contrast to the predominant belief of his day that women's education should be limited (at the very most) to basic literacy and practical skills. He supported the idea that women had a right to enjoy intellectual study as much as men did.

Thomas Sprat (1635–1713) took up Comenius's ideas in his book *The History of the Royal Society of London, for the Improving of Natural Knowledge* (1667). Sprat believed that the new developments in science were not only expanding human understanding of the natural environment, but also forging a new cultural and literary tradition (Phillips 1990, p. 28): scholars were beginning to write in English and other vernacular languages, instead of using Latin, in the hope of securing greater public exposure for, and understanding of, their achievements. Sprat does not specifically include women in his vision, but the use of English, as advocated by Sprat and others, undoubtedly made scientific writings more accessible to those women who could read but had not received any instruction in Latin.

WOMEN ADVOCATING
EDUCATIONAL REFORM

This section presents the work of three women who became prominent in the movement for educational reform: Anna Maria van Schurmann, Bathsua Makin, and Mary Astell.

It should be noted, however, that there were other women pursuing similar aims about whom information is now lacking.

Margaret Alic (1986, p. 78) has described Anna Maria van Schurmann (1607–1678) as one of the first feminists to speak out for women's scientific education, and Patricia Phillips (1990, p. 21) remarks that Van Schurmann's name became the one most often quoted by men who wrote on women's matters in her lifetime. Van Schurmann was born of a Dutch father and a German mother. She received her early education at home, being tutored with her brothers in languages and the classics. She also became knowledgeable in mathematics, calculus, and astronomy, and versed in poetry, rhetoric, dialectics, and philosophy. As the best Latinist in the Dutch city of Utrecht, she was asked in 1636 to write verses for the inauguration of its new university, and later she was given the rare privilege of attending lectures there, although she was required to sit behind a curtain while doing so. In 1639, she published her *Dissertatio, de ingenii mulieribus ad doctrinam, et meliores litteras aptitudine*, an essay on the capacity and aptitude of women to learn, and in 1659 she wrote a book that became known in its influential English translation as *The Learned Maid, or Whether a Maid may be a Scholar* (see Disse 2009).

Van Schurmann's approach was that girls should be provided with the same classical education as boys. During her active years, she had a positive influence on other women who were also working towards a similar goal, and her correspondents included Marie de Gournay, to whom she wrote arguing that women should study science, although she appears to have meant only single women rather than married women (Waters 2000, p. 43). Her final years were spent in a religious group, where she abandoned her classical studies to concentrate on the Scriptures.

Another of Anna Maria van Schurmann's correspondents was the English schoolteacher and writer Bathsua Makin (c. 1600–c. 1675), about whom less is known: her biographer

Frances Teague mentions that, for example, "it is not known whether her school was a success, how long Makin taught there, or even when Makin died" (1998:104). At the age of sixteen, Bathsua had written verses for King James I and his family, and she had begun soon after to tutor some of the King's daughters, even though (as mentioned above) James was not supportive of education for women. She married Richard Makin in 1620 and had at least nine children in twenty years, six of whom survived. With her husband away from home for years at a time, she was financially responsible for providing clothing, food, medicine, and other household needs. Among her achievements was the development of a system known as "radiography," a form of shorthand allowing its users to take notes quickly in several languages. At some point she came into contact with the work of Comenius through her sister's husband, the mathematician John Pell (1610–1685), and she opened her school for girls at Tottenham High Cross, near London, in 1673, the same year in which she published her major work, *An Essay to Revive the Antient Education of Gentlewomen*, using a male pseudonym (as many other women did well into the 19th century).

The *Essay* begins with a prefatory letter addressed to women (Makin 1673, p. 1), stating that they need education as much as men do. Makin concedes that many people will oppose this position, but she argues that educated women will make better wives. She also contends that educated women will make the next generation of men more industrious in *their* studies. In closing, Makin points out that educated women will be able to help husbands with advice, but she is careful to add that husbands will retain the casting vote in all decisions. A second letter (Makin, 1673, p. 2) makes the point that educating women will benefit men as well. In a third letter, Makin pretends to be a man who despises the idea of educating women and puts forward arguments against doing so, but her fourth letter refutes the arguments of the third letter point by point. (Makin, 1673, pp. 2–3). The essay itself provides examples of learned women who are eminent in arts and languages, are good orators, profound philosophers or poets, and understand logic and mathematics. (Makin 1673, pp. 3–15). Makin profiles these educated women to refute prejudices about women's abilities to learn. Later in the essay she makes the point that education will benefit every woman, no matter what her personal situation may be, whether maid, wife, or widow. The last part of the pamphlet is an advertisement for her school.

The strategy Makin developed for her school was to market her educational approach as traditional and aimed at parents who did not wish their girls to receive more than basic skills, even though in actuality she taught grammar, rhetoric, logic, physic (basic medicine), languages (especially Hebrew and Greek, for better understanding of the Scriptures), mathematics and geography, history, painting, and poetry, in addition to sowing and music (Teague 1998, pp. 130–31). Like Marie de Gournay and Mary Ward, Bathsua Makin valued education that would allow "women to earn their own living, manage their own affairs, and defend their own homes" (Teague 1998, p. 88). She viewed an educated woman as one who could act for herself and be independent. This belief, coupled with the teaching of science and mathematics to girls, amounted to quite a revolutionary approach for the period. Her school for girls, and some other girls' schools too, followed this programme for almost two hundred years, and provided a model for some of the boys' schools reforms in England in the mid-19th century.

The trail blazed by Bathsua Makin was followed by Mary Astell (whose views we have already discussed in Chapter 2). She was born into a family of merchants but lost both her parents while she was in her teens and went to live in the household of Lady Catherine Jones. There she came into contact with a number of well-educated women who were interested in changing the status of women. In 1696, Astell published her first work, *A Serious Proposal to the Ladies for the Advancement of Their True and Greatest Interest.* In this work she encourages women to develop their minds and instructs them how to think clearly and logically. However, like Van Schurmann, she saw her main objective not as encouraging women to take public positions or usurp male authority but as enabling them to be better Christians. Thus, she makes it clear (as quoted in Waters 2000, p. 43),

> We pretend not that Women shou'd teach in the Church, or usurp Authority where it is not allow'd them; permit us only to understand our *own* duty, and not be forc'd to take it upon trust from others; to be at least so far learned, as to be able to form in our minds a true Idea of Christianity, it being so very necessary to fence us against the danger of these *last* and *perilous* days. . . . And let us also acquire a true Practical Knowledge such as will convince us of the absolute necessity of *Holy Living* as well as of *Right Believing*, and that no Heresy is more dangerous than that of an ungodly and wicked Life.

As for the question of how men would react to the advice she was offering to women, Astell responds (again, as quoted in Waters 2000, p. 43),

> The Ladies, I'm sure, have no reason to dislike this Proposal, but I know not how the Men will resent it to have their enclosure broke down, and Women invited to tast[e] of that Tree of Knowledge they have so long unjustly Monopoliz'd. But they must excuse me, if I be as partial to my own Sex as they are to theirs, and think Women as capable of Learning as Men are, and that it becomes them as well.

Mary Astell also wrote an essay on marriage, in which she criticizes men who marry merely for money or because of the beauty of their wives, and advises women not to marry out of a sense of duty, or to escape hardship, but to base such an important decision on reason. She reminds women that when they marry, they become "upper servants" to men, and will have to obey and be submissive wives, itself a major reason to choose their spouse carefully. Astell reminds women that they can only refuse an offer of marriage, not make one, and expresses the view that marriage seems to be more of an advantage for men than for women, although the human race would come to an end if women did not marry (Waters 2000, p. 43).

In this overview of women's education in the early modern era, and its connections with their limited social roles and vocations, it has been interesting to observe that some girls' schools taught science and mathematics, in addition to many other subjects. This strategy, promoted by Comenius and Bathsua Makin among others, provided a universal education for those girls who could afford to attend such schools. It took another two hundred years for school reforms to add science and mathematics to the curriculum in boys' schools, but these reforms also brought some negative changes to the curriculum for girls (as will be seen in Chapter 7). Nevertheless, Comenius, Makin, Van Schurmann, Astell, and their colleagues did succeed in introducing new ideas and providing new learning opportunities for girls, and it is regrettable that it has taken so long to rediscover their work.

———⊷∞∞⊶———

CHAPTER 6

Education for Women
in the 18ᵗʰ Century

By the 19ᵗʰ century, the power centre of science shifted once again as the academies and other learned societies became less prominent and were replaced by other types of institutions. However, academies and scientific societies still played a major role in the 18ᵗʰ century and continued to exclude women. James E. McLellan (1985, pp. xxi–xxii) summarizes what happened:

> The 18ᵗʰ century was the heyday of the general scientific society. By the 19ᵗʰ century learned societies ceased to be the premier institutions for the organization and pursuit of science. That role was taken over by specialized professional scientific societies, and by university-based teaching and research. . . . The 18ᵗʰ-century type of scientific society (academy) survives as a local or provincial social group, as an honorary organization to which one is elected at the end of an active scientific career, or as an overarching bureaucratic entity controlling the research efforts of subordinate units. . . . In the 18ᵗʰ century, the scientific enterprise grew considerably

larger and became better integrated into society. . . . By 1800, the social profile of science had changed considerably . . . [with] many more serious practitioners of science . . . [and] many more niches in society where scientific knowledge was valuable and where the man of science could make a home. Governments at all levels had become convinced that science was useful to the state, and they had incorporated scientific and technical expertise into their service. . . . By the 19th century science had become tightly woven into the cultural fabric of the West.

When science was still relatively poorly regarded by society, in the 16th and 17th centuries, women had found fewer obstacles to studying or practising science, as we have seen exemplified (in Chapter 4) by the numerous women astronomers of the era. As the enterprise of science was reorganized into formal bodies that excluded women, it became more difficult for women to engage in these activities, the main obstacle being their lack of access to scientific education and to the instruments necessary for the practice of science, from telescopes and microscopes to such equipment as the vacuum jar. Women also had difficulty finding publishers who would agree to print and publish their writings, which is why so many adopted pseudonyms. In spite of these and other obstacles, some women found ways to study science and mathematics, to get involved in scientific work, to write, and even to get published. The common factors that enabled these women to achieve some measure of success and recognition were their genuine interest in, and curiosity about, science or mathematics; their abilities in these subjects; and, almost always, support from men who encouraged them to pursue their interests in spite of social convention. We can only imagine what these women could have achieved if they had been allowed free rein.

DIVERGENT VIEWS OF WOMEN'S ABILITIES

The common belief in the inferiority of women became even more entrenched in 18th-century Europe than it had been in the 17th century. As we saw in Chapter 2, Rousseau, Hume, Kant, Goethe, von Humboldt, and other influential thinkers believed that girls and boys should receive completely different types of education. What is interesting is that several women writers of the day agreed with them. This section presents a selection of their views, and the views of their opponents, drawing on the work of Phyllis Stock (1978, pp. 109–15).

Anne d'Aubourg de la Bove, comtesse de Miremont (1735–1811), wrote a seven-volume course on girls' education (1779–89), which Stock summarizes as follows: "Women were not destined to learn anything in depth. The study of religion and the accomplishments was to be enriched by the three Rs, [that is, reading, writing, and arithmetic,] grammar, geography, history, and natural science ... [but] women should never appear learned."

Hannah More (1745–1833) argued in her *Essays on Various Subjects, Principally Designed for Young Ladies* (1777) that moral reform should start with the aristocracy as role models for the young and the lower classes. As one of the most successful writers of her day, More used her huge influence to recommend that women "restrain themselves, to give up public expression of their own opinions, even if they were right."

Stéphanie Félicité du Crest de Saint-Aubin, comtesse de Genlis and marquise de Sillery (1746–1830), wrote *Adèle et Théodore* (1782), a novel that became popular in England as well as in France, and in which she described the proper education for a girl and for a boy. The girl's day consists of religious observances, household duties, reading and reciting, drawing, music, and counting. In the meantime, the boy, under the guidance of his father, studies Latin, reads books on law and politics, and reads more books while his sister sews. He does not study music.

Stock also presents the views of the physician P. J. G. Cabanis (1757–1808), which are representative of the mainstream medical opinion of the time. Cabanis "opposed public education for girls on the basis of their weakness," because he believed that it "was dangerous to expose women to the perils of a life that their constitution could not support without being denatured." As in earlier centuries, women were widely held to be subject to biology: in effect, the nature–nurture debate was settled before it could even begin.

However, even in the first half of the 18th century, the defence of women's rights to education and public positions was taken u₁ by a few women, and some men as well. For example, in 1739 an author known only as "Sophia" (the ancient Greek for "wisdom") published an essay, *Woman Not Inferior to Man*, in which she argued that, as Stock summarizes it,

> women were slaves, who had achieved less than men only because they had been given less education. . . . ["Sophia"] argued that women should be allowed an independent role in society. They could become doctors, lawyers, teachers, even soldiers, if they were not limited to household duties and were given a good education.

"Sophia" made little impact, but in the latter part of the 18th century advocates for the education of girls, including Mary Wollstonecraft (discussed in Chapter 3), defended the concept of universal education. In France, the mathematician and pioneering social scientist Marie Jean Antoine Nicolas de Caritat, marquis de Condorcet (1743–1794), advocated universal education for the rich and the poor, and for girls and boys alike, and, in his *Lettres d'un Bourgeois de New Haven à un Citoyen de Virginie* (1787), he also argued that all professions should be open to both sexes. Condorcet believed that a similar education should be provided to women and to men, to prevent inequality between the sexes in the family, but also for the sake of simple justice. He felt that women had the same right to public education as men did. Later, supporters of the concept of popular, universal education used some of Condorcet's arguments, but it took almost one hundred years to rediscover his specific reasoning on the education of women. The French Revolution, whose banner proclaimed "Liberty, Equality, Fraternity," did not apply these principles equally to men and women in France. This can be seen as a missed opportunity to bring equality to all citizens.

In England, at around the same period, Catherine Macaulay (1731–1791), a writer and historian, published her *Letters on Education* (1790), in which she dismissed accepted views on the inferiority of women (see Macaulay 1994). She recommended that girls and boys be educated together, do the same physical exercises, and study the same school subjects. She demanded political rights for women, and wanted women to use their education and talents to win in a man's world, just as she felt she had.

WOMEN'S INFORMAL WAYS OF STUDYING AND PRACTISING SCIENCE

It did not help women that during the 18th century the convent schools and boarding schools that provided formal education for girls were beginning to acquire a bad reputation. Research by Carolyn Lougee (1999, pp. 193, 208, 210) shows that these institutions were increasingly notorious for the poor quality and inadequate portions of the food they provided, the excessive rigour of the religious devotions they required, and their high levels of contagion, internally generated infections, and progressive debilitation: in France, for example, mortality rates among children at boarding schools were higher than the average for the same age group in the population as a whole.

If girls were to receive any education at all, it was preferable to arrange for it to be provided at home by male relatives or tutors, and closely supervised by their mothers.

In spite of this continuing lack of access to formal education, some women found ways to acquire scientific knowledge and explore ways of becoming involved in scientific activities. Many of them were encouraged by a male figure in their families, whether their fathers, their brothers, or their husbands. The exclusion of women from the new scientific centres of power also stimulated the creation of resources with which to break down the barriers and achieve higher scientific learning.

As scientific activity became increasingly regulated and exclusive, the publication of scientific works came under the direct control of the academies and other learned societies. Specialization in subdisciplines was also another social construction of science that contributed to its formalization. However, it was still possible for amateurs and hobbyists to practise science, at least for people of means, since to carry out experiments and write on scientific topics one needed basic training in science and mathematics, as well as access to a library, a laboratory, and instruments. In the West Midlands of England, for example, where a group of fourteen progressive and inventive men created the Lunar Society—holding monthly meetings on the Monday closest to a full moon in order to be able to travel more safely after dark—the members' wives often participated in their discussions and became known as the "lunar ladies" (Phillips 1990, p. 163). Another group of women in the Black Country arranged science evenings, with observations of the sky, experiments in physics, chemistry, and biology, and discussions of nature and philosophy. The fame achieved by the female astronomer Caroline Herschel (discussed below) also encouraged many women to observe the sky during social evenings, especially in Bath, London, and other centres of fashion.

Scientific tourism was another means of informal education, in which women visited factories and practised fieldwork. One of the difficulties they encountered was the cumbersomeness of their clothing when they were climbing down into caves or along cliffs in search of geological or pale ontological specimens. Some of these women were self-taught and tutored others in their families. Others established schools, in the tradition of Bathsua Makin's, where science played an important part in the curriculum (Phillips 1990, pp. 137–138). A few of these women were truly adventurous: Maria Sybilla Merien (1647– 1717) spent two years travelling through Surinam, in South America, in pursuit of her study of insects (Schiebinger 1989, p. 74).

Another informal approach to studying science was to read textbooks and journals, especially those intended for women readers and cheap reprints or adaptations of books from the late 17ᵗʰ century. The first book on a scientific subject written for a general audience was probably William Leybourn and Vincent Wing's *Urania Practica—Rules and Astronomical Tables* (1648), though Robert Boyle's *New Experiments* (1660), which provided descriptions of the demonstrations he performed during his lectures at the Royal Society, had a wider appeal. At a time when most scholarly works were still written in Latin, Boyle believed in making science understandable to anyone who could read English. His example was followed by Henry Power, author of a textbook on *Experimental Philosophy* (1664), and by Robert Hooke, whose *Micrographia* (1665) introduced a wider readership to the marvels of microscopy. These books, which remained popular for decades, paved the way for a whole industry producing do-it-yourself books and guides to elementary science, as well as flash cards and games introducing geography, geometry, and astronomy to amateurs. Instruments such as globes, microscopes, and telescopes became available at much lower prices than before, and thus became more accessible to people of lesser means.

In France, meanwhile, a new style of scientific writing
for a popular audience was inaugurated by Bernard Bovier,
sieur de Fontenelle, in his *Entretiens sur la pluralité des
mondes* (1686, *Conversations on the Plurality of Worlds*). In
French rather than in Latin, Fontenelle discusses astronomy,
physics, and microscopy within the structure of a series of
conversations between a young male teacher and his pupil,
a *marquise*. Fontenelle's use of the conversational style was
imitated by numerous authors, notably the Italian Francesco
Algarotti (1712–1764), whose book *Il Newtonianismo per le
dame* (1737) explained Sir Isaac Newton's ideas to women
readers. The best-known scientific writer of the age was
probably George-Louis Leclerc, comte de Buffon, whose
massive *Histoire naturelle, générale et particulière* (1749–88),
in forty-four volumes, popularized the sciences of botany
and zoology, attempted to describe everything that was then
known about the natural world and included a discussion of
the similarities between humans and apes one hundred years
before Charles Darwin.

Scientific journalism also took on new forms during this
period. The Royal Society in London had published its weekly
Philosophical Transactions ever since 1665, and in 1682 it
began issuing *Weekly Memorials for the Ingenious, or An
Account of Books Lately Set Forth in Several Languages,
with Some Other Serious Novelties Relating to Arts and
Sciences* (Phillips 1990, pp. 82–84). Publishers brought out
similar digests of the latest scientific ideas and discoveries
in response to a growing audience of informed readers, both
male and female. These included *The Lady's Diary*, which
first appeared in 1704 and continued until 1840, which
focused on scientific topics and presented mathematical
problems to which hundreds of women responded every year;
The Spectator, which included reports on science alongside
its coverage of politics and literature; the *Female Spectator*,
which attempted the same broad coverage, but lasted only
from 1744 to 1746; and, during the closing years of the

century, the *Lady's Monthly Museum*. As in previous centuries, these periodicals claimed to promote a higher moral tone for women, presenting their scientific content in forms that did not openly challenge social conventions.

The salons (already mentioned in Chapter 4) offered another venue for women seeking to study and participate in science. Although the salons were far more popular in France than they were in England, there were a few in London that exerted considerable influence among the upper classes, including Mrs. Vesey's and Mrs. Thrale's. Mrs. Vesey was well-known as one of the "bluestocking ladies," who generally studied the classics and had little interest in science, apart from the "respectable" fields of astronomy and mathematics. Mrs. Vesey was exceptional in taking an active interest in other scientific topics. Another "bluestocking lady" with an interest in science was Elizabeth Carter (1717–1806), who gained an international reputation as a classical scholar and linguist, as well as for her work in mathematics and astronomy. She was considered the leading intellectual woman in England in her time. Her book, *Sir Isaac Newton's Philosophy Explained for the Use of the Ladies in Six Dialogues on Light and Colour* (1739) was more than a translation of Francesco Algarotti's book (mentioned above), for it clarified and simplified the scientific and philosophical concepts in Newton's *Principia Mathematica*, rendering them more accessible to readers (Phillips 1990, p. 92).

EXAMPLES OF WOMEN WHO PRACTISED SCIENCE AND MATHEMATICS IN THE 18ᵗʰ CENTURY

This section profiles some of the women who persevered in their determination to practise science or mathematics, despite their exclusion from the universities and the academies. The men who encouraged these women should also be remembered, because their support went against the social conventions of the times.

Some female members of the royal families of Europe were notable for their support of science. Caroline of Ansbach, the wife of King George II of England, organized evenings of scientific discussion at her husband's palace, and her son Frederick, Prince of Wales, also had an interest in science. His wife, Augusta, who had been tutored by the botanist and chemist Stephen Hales (1677–1761), helped in the founding of the Royal Botanical Gardens at Kew, near London.

Italy was unique among European countries in the early modern era in allowing women, albeit only in small numbers and only from wealthy or noble families, to study and teach in universities. The University of Bologna had allowed women to attend lectures from its inception, in 1088, and Dorotea Bucca had occupied a chair in medicine there in the 15th century. An outstanding example of a woman who held a professorship there in the 18th century is Laura Maria Caterina Bassi (1711–1778) (see Nies 1999). After being educated at home by her family's physician, Gaetano Tacconi, she impressed the city's professors and learned gentlemen with her knowledge of Cartesian and Newtonian philosophy, and in 1732 she was appointed Professor of Anatomy. Two years later, she was given a position in philosophy and obtained permission to expand her role by offering private lessons in experimental physics, introducing Newton's theories of light and optics, and the laws of motion he had formulated. She and her students performed experiments on electromagnetism in her own laboratory, and she published twenty-eight papers (thirteen on physics, eleven on hydraulics, one on mechanics, one on chemistry, and two on mathematics), copies of which can still be found in the Academy of Science in Bologna. In 1738, she married a fellow scientist, Giovanni Giuseppe Veratti, who became her assistant. Together they investigated the medical uses of electricity, pioneering the modern field of biomedical engineering. Laura Bassi had eight children, five of whom survived, yet she managed to find time to encourage the scientific careers of her cousin Lazzaro Spallanzani and of other younger scientists, such as Alessandro Volta. By 1760, her annual salary of 1,200 *lire* was higher than any other professor in the scientific departments at the university, and was even more than the president of the university earned. In 1776, when Laura Bassi was sixty-five years old, she was appointed to a chair in physics at the Institute of Sciences in Bologna, replacing Paolo Balbi. Her husband became her teaching assistant and, after she died two years later, he replaced her in this position.

Unfortunately, Laura Bassi's example did not open doors for other women, because the male scientists and decision-makers of Bologna were reluctant to allow another woman the latitude they had provided to her (Wertheim 1995, p. 139). The men who supported her career were her father, who believed in her education, her tutor Tacconi, her husband, who lobbied for her to get the chair in physics, and especially Prospero Lorenzo Lambertini, who became Pope Benedict XIV, an important patron who is said to have ordered the university to grant her a degree.

In 1751, the University of Bologna awarded a degree to another woman, Cristina Roccati (1732–1797), who is believed to have been only the third woman to obtain a university degree anywhere in Europe. Cristina Roccati seemed to be on her way to becoming another Laura Bassi, but her family was financially ruined and she moved to Rovigo, a provincial town, where she taught physics at the local scientific institute for twenty-seven years. Her complete lectures on Newtonian physics survive in manuscript form (Wertheim 1995, p. 140).

Another outstanding Italian woman was Maria Gaetana Agnesi (1718–1799) (see Kennedy 1987). The first of the twenty-one children of the first of her father's three wives, she was reportedly writing, reading, and speaking seven languages (Italian, French, Latin, Greek, German, Spanish, and Hebrew) by the age of nine. Her father encouraged her to take part in debates on philosophy and science with him and his colleagues, and she learned mathematics from tutors who later became professors. In 1748, she published *Instituzioni Analitiche*, a work in two volumes on algebra, analytic geometry, calculus, and differential equations that helped to establish her fame. Maria Agnesi was appointed as a member of the Academy of Sciences in Bologna, and Pope Benedict XIV also appointed her as an honorary professor at the University of Bologna in 1750. Although she was urged by Laura Bassi and others to accept this position, she chose instead to devote the rest of her life to helping the poor and the sick.

Another extraordinary woman of this time was Gabrielle-Émilie Le Tonnelier de Breteuil, marquise du Châtelet (1706–1749), a mathematician, philosopher, and physicist (see Tee 1987). Her mother was the daughter of a French nobleman, and her father was a baron and chief of protocol at the French court. When her father noticed her obvious academic talents, he hired tutors to teach her Latin, Italian, and English, but mathematics was her favourite subject. After her marriage, at the age of nineteen, to the thirty-year-old marquis du Châtelet, she hired her own tutors and focused on studying mathematics. She had three children in the early years of her marriage, but this did not curtail her studies: like other wealthy educated women, she was enabled to pursue her studies by the work of wetnurses and governesses, who relieved her of the burden of child-rearing. In 1730 she met the philosopher Voltaire (1694–1778), who had recently returned to France after spending three years in England, and in 1734 they moved to her husband's estate in Cirey en Champagne, where they lived together while her husband was away on military duty. It was during this period that she produced her principal works. The first was a joint publication with Voltaire, *Éléments de la philosophie de Newton* (1739), but a year later she published her own book, *Institutions de physique*, a detailed and clear exposition of Leibniz's physics. From 1745 she concentrated on her other major work, a translation of

Newton's *Principia* from Latin into French. This was a major accomplishment, requiring a thorough understanding of this difficult treatise. Her main objective was to make it available to scientists in France in their own language, but the work also contained additions and transpositions of ideas, and thus went beyond being a direct translation. The book was written at a time when Cartesian philosophy was dominant in France and Newton had few supporters there. When she was forty-three, Émilie gave birth to a fourth child who is believed to have been fathered by her last lover, the marquis de Saint Lambert. The child was born while she was writing her last book. She seemed well after the birth, but died quite suddenly one week later. The child also died soon after birth. In 1759, Voltaire ensured the posthumous publication of her last book, which is still the only French translation of Newton's *Principia*. In spite of her great achievements, Émilie du Chatelet's substantive work was, like that of many other women who lived before and after her, largely ignored. Whenever she appears in traditional historical narratives, it is usually her liaison with Voltaire that justifies her inclusion, not her own deep understanding of Newton's natural philosophy and Leibniz's metaphysics, or even her own additions to their works.

Women continued to work in astronomy during the 18th century and their accomplishments were remarkable. Nichole-Reine Lepaute (1723–1788), an assistant to the French astronomer Joseph Lalande, successfully predicted the return of Halley's Comet in 1759 by calculating the gravitational effects of Jupiter and Saturn on the comet's orbit. She also predicted an annular eclipse of the sun that would be visible in France and created a chart for the whole of Europe, indicating the time and type of eclipse, whether full or partial. Lepaute married a royal clockmaker and provided him with calculations of the number of oscillations per unit of time for pendulums of various lengths. The results of her work were published in a *Traité d'horlogerie* under her husband's name.

The most famous woman astronomer of the day was Caroline Lucretia Herschel (1750–1848), who was born in Hanover in Germany, one of ten children (see Kronk 2008). Her mother did not support the idea of her receiving a formal education, but her father secretly provided her with the means to study music and mathematics, with the support of her older brother Friedrich Wilhelm (later William, 1738–1822). In 1756 William moved to England and in 1772 he invited Caroline to join him there as his assistant. He began to improve the design of telescopes, discovered the planet Uranus in 1781, and the following year was appointed astronomer to King George III. On every clear night, Caroline finalized the observations of the day and planned the next schedule of observations. The Herschels found the first object for their catalogue of nebulas and star clusters in 1782, and built a larger reflector the following year. It was while her brother was travelling in Germany in 1783 that Caroline found a comet moving slowly through Leo. This discovery was confirmed by several European astronomers to whom she sent her findings. In the same year she discovered an open cluster of stars and a galaxy (the astronomer Messier had noticed the galaxy in 1773, but had not included it in his list of celestial objects). In 1787 she began to receive a royal salary of fifty pounds a year so that she could continue as William's assistant, making her the first woman to be provided with an official scientific position in England. After William's marriage, in 1788, Caroline continued to work as his assistant, but eventually began to work independently. In 1788, she discovered her second comet, now known as the Herschel–Rigollet comet (because Rigollet made observations of it when it returned in 1939). Between 1786 and 1797 Caroline discovered eight more comets and three nebulas (Schiebinger 1989, p. 263). In 1798 she submitted to the Royal Society a star catalogue that contained 560 more stars than Flamsteed's standard catalogue did.

After her brother William's death, in 1822, Caroline returned to Hanover and completed his catalogue of 2,500 nebulas, for which she received the Gold Medal of the Royal Astronomical Society in 1828. She was the first woman to receive honorary membership of the Royal Society, in 1835, was elected to the Royal Irish Academy in 1838, and received a Gold Medal for Science from the King of Prussia in 1846. In 1889 a minor planet was named "Lucretia" in her honour.

— ⚬⚬⚬ —

—∽∾∿∾∽—

CHAPTER 7

School and University Reforms in the 19th Century

The traditional ideas discussed in previous chapters retained their dominance throughout much of the 19th century, notably, that women were, by nature, not fit to study, and might even fall ill or die from the strain of acquiring an education; and that those few women who could benefit from formal education would lose their femininity in doing so, and become less desirable as wives. These ideas were propagated not only by men, but, as in previous centuries, by many mothers who believed that their daughters should concentrate on preparing for marriage rather than risk their health or their social status by seeking education. However, it was in the 19th century that the counterarguments in favour of education for women began, at long last, to make an impact on western societies, with the founding of more and more schools for girls and of the first institutions of higher education for women.

SCHOOL REFORM IN ENGLAND

Between 1864 and 1868, a royal commission, known by the name of its chairman as the Taunton Commission, examined the provision of formal education for boys and girls in both "endowed schools" (the Victorian term for schools that where wholly or partly funded by public money) and "proprietary schools" (privately owned establishments, also known as "independent schools"). Its report tells us a great deal about the state of formal education in England at that time, as well as the prevailing attitudes to education.

Interestingly, in their chapter on girls' schools (see Schools Inquiry Commission 1867–68), the commissioners directly address the social prejudices that shaped the education of girls at that time. They begin by quoting Ralph Lingen, the chief official of the government's Education Office:

> If one looks to the enormous number of unmarried women in the middle class, who have to earn their own bread, at the great drain of the male population of this country for the army, for India, and for the colonies, at the expensiveness of living here, and consequent lateness of marriage, it seems to me that the instruction of the girls of a middle-class family, for any one who thinks much of it, is important to the very last degree.

They then point out,

> We have had much evidence showing the general indifference of parents to girls' education, both in itself and as compared to that of boys . . . There is a long-established and inveterate prejudice, though it may not often be distinctly expressed, that girls are less capable of mental cultivation, and less in need of it, than boys; that accomplishments, and what is showy and superficially attractive, are what is really essential for them; and, in

particular, that as regards their relations to the other sex and the probabilities of marriage, more solid attainments are actually disadvantageous rather than the reverse.

On the one hand, they respond by agreeing with such parents, to an extent that now seems startling but was then conventional:

Such ideas . . . have a very strong root in human nature, and with respect to the average, nay, to the great majority of mankind, it would be idle to suppose that they would ever cease to have a powerful operation. Parents who have daughters will always look to their being provided for in marriage, will always believe that the gentler graces and winning qualities of character will be their best passports to marriage, and will always expect their husbands to take on themselves the intellectual toil and the active exertions needed for the support of the family.

On the other hand, they then challenge such prejudices, apparently without noticing any contradiction:

There is weighty evidence to the effect that the essential capacity for learning is the same, or nearly the same, in the two sexes. This is the universal and undoubting belief . . . throughout the United States; and it is affirmed, both generally and in respect to several of the most crucial subjects, by many of our best authorities. . . . [Also,] on the special point of the health of women, both in youth and in after-life, . . . so far from its being true that they are likely to suffer from increased and more systematic intellectual exercise and attainment, the very opposite view is maintained, both as the result of experience and on scientific authority.

Overall, the commission's report revealed poor levels of provision for boys and girls alike, especially in secondary schools (those for children aged eleven and above), and they criticized the uneven distribution of public endowments, which varied from place to place, depending on accidents of history, and were often misused. The commissioners also noted that there were only thirteen secondary schools for girls in the whole of England, and that girls' schools had the lowest standards of all. Most of the girls' schools taught only the traditional "accomplishments" (referred to above) of music, needlework, and conversational French, although the Quaker schools were singled out for commendation for providing equal education for both sexes and a strong curriculum, including experimental science, and a few girls' schools were also reported as integrating science into the curriculum (Phillips 1990, p. 244). For example, at the North London Collegiate School, an independent school founded in 1850, the headmistress, Frances Buss, had introduced a curriculum that included topics such as "the property of matter, the laws of motion, mechanical powers, simple chemistry, electricity, geology, botany, natural history and astronomy" (Phillips 1990, p. 246). However, as Patricia Phillips (1990, p. 251) observes,

> For all their enthusiasm and approval for the modern and scientific education they had observed in the best of the girls' schools, the assistant commissioners were influenced in their final conclusions by their own unavoidable educational prejudices. With the best will in the world, they suggested that the way in which girls' schools might be improved was to increase the quota of classics taught there.

The controversy over the Taunton Commission's findings led to the creation in 1869 of the Endowed Schools Commission, which drew up new governing structures for these schools and attempted to impose uniform rules for the distribution of public funds (Phillips 1990, pp. 237–39). In 1870 elementary education was made compulsory for all boys and girls up to the age of eleven (it was only in 1944 that fees for secondary schools were abolished and they became accessible to all).

The Taunton Commission was followed by the Bryce Commission of 1895, which again surveyed the state of schools in England. This time, the commissioners concluded that the female teachers in girls' schools were "not fully equal to their task," for they had not been well taught themselves and did not know how to teach. According to the commissioners, this was the main reason why girls' schools still achieved poor results compared to boys' schools.

The movement for reform that led to the creation of these two royal commissions also raised the issue of coeducation: should girls be educated in the same schools as boys, or separately? Admitting girls to boys' school, as was done in Belgium, Germany, and Italy, enabled them to get a better education, while in the United Kingdom and France all schools remained exclusively single-sex, and even those with mixed intakes had separate classes for boys and girls. In contrast, the United States and Canada had both coeducational and single-sex schools and colleges. One main argument in favour of coeducation, where it was supported, was financial rather than pedagogic: having one building meeting the needs of both girls and boys, especially in remote areas with sparse populations—which, of course, were much more prevalent in the United States and Canada than in western Europe—was more economical than doubling the number of buildings.

The Langham Place Group was an association of women that demanded a major reshaping of secondary schools for girls in England. While its leader, Barbara Leigh Smith Bodichon (1827–1891), founded a coeducational school, Portman Hall, in 1854, and visited coeducational schools in the United States in the late 1850s, the group did not advocate coeducation among their demands for reform. The university reformer Emily Davies (see below) also supported the model of separate schools for girls: in January 1868, she admitted that she would have preferred coeducational institutions, but she considered them unattainable in England (Albisetti 2000, p. 476). Other women educators who supported coeducation education were Elizabeth Wolstenholme, Josephine Butler (for economic reasons), and Mary Grey. In an address delivered in 1871, which led to the establishment of the Women's Education Union (WEU), Mary Grey said (as quoted in (Albisetti 2000, pp. 476–77),

> There is much and weighty evidence to show that it would be an advantage to both sexes, morally as well as intellectually, as well as an immense saving of money and teaching power, that they should climb it together from the infant school to college inclusive, learning by this common work for common aims to recognize their common human nature, to complement what in each is deficient, and thus to become the true helpmates which God created them to be. The time, however, may not be ripe for this yet.

There is no doubt that the work of the Taunton and Bryce commissions, and the efforts of the women educators they encouraged, had a major impact, helping to extend the formal education of girls from the elementary level to secondary schools and even to colleges. However, the teaching of science and mathematics to girls tended to dwindle from the 1860s, at the same time as the new curriculum introduced in many schools for boys included more science. This divergence has been perpetuated ever since (not only, of course, in England), so that today it is still common to observe that high school physics is a subject where the majority of students are boys. The same can be said of courses on computers and information technology.

EMILY DAVIES AND UNIVERSITY REFORM IN ENGLAND

One of the main figures behind university reform in England was Emily Davies (1830–1921). Born in Southampton, Emily was the second daughter and fourth child of a cleric, Dr. John Davies, and his wife Mary Hopkinson Davies, the daughter of a well-to-do businessman. Emily received a very limited education at home and was not allowed to join her brothers, who were being prepared by tutors to attend boarding school and then university. She felt herself "imprisoned" in the family home for the first thirty years of her life. It is quite remarkable that, in spite of her own limited education, she achieved major gains for women's access to higher education in the space of a few years.

In the 1850s, Emily Davies campaigned for women to be admitted as students at University College, London, in support of her friend Elizabeth Garrett, who wanted to study medicine there. The plan did not succeed, in spite of a petition signed by 1,500 eminent men supporting the campaign: George Grote, the vice-chancellor of the college, even scrutinized the signatures with a magnifying glass and accused Emily Davies of forging many of them. This humiliation was followed by a majority vote in the Senate of the University of London (of which University College is a member body), denying admission to women. Elizabeth Garrett was finally admitted to the medical schools of the Middlesex Hospital and the London Hospital for some of her training, but she was treated with scorn and felt isolated among the rowdy male students. She was eventually accepted at St. Andrew's University in Scotland. Emily Davies frequently had to encourage her during her many periods of discouragement as she struggled to become the first woman in the United Kingdom to graduate as a doctor.

Emily Davies then turned her attention to a successful campaign to allow girls to enter the new school examinations, sponsored by the universities, on equal terms with boys, and in 1864 she became the first woman ever to give evidence to a royal commission (Bennett, p. 1). It was partly due to her evidence that the Schools Enquiry Commission included in its report the assessment of gender inequalities that is quoted from above (Phillips 1990, p. 236). Not content with theorizing, she then formed a group of women to raise money to purchase a house at Hitchin in Hertfordshire—which was as close as she could get to Cambridge—and there, in 1873, they opened Girton College. Girton was the first institution in the United Kingdom dedicated to the higher education of women, and, as Davies's biographer Daphne Bennett (1990, p. 1) explains, it "was soon an accepted feature of Cambridge life, but the obstinacy of a masculine and still almost monastic society prevented her from attaining admission to full membership of the university and to its degrees." It was not until 1948 that students from Girton and the other women's colleges founded afterwards in Cambridge could finally receive degrees from the university.

Emily Davies took on huge tasks as she sought to persuade men to open the doors of formal education to girls, especially at the level of secondary school and university, and to convince mothers of the importance of education and economic independence for their daughters. Although she did not win all her battles against the patriarchal system, she played a decisive role in initiating higher education for women in England, and also fought for the right of women to vote, which was finally achieved in 1920, just one year before her death.

One concern about Emily Davies's work is that it tended to steer women towards a conservative view of education, following the line of what men had valued in the past. This, unfortunately, did not include science. Since Latin was then a prerequisite for attending university, and men concentrated on the study of that language in the independent schools and some of the State schools, Davies concluded that women needed to follow the same pattern in order to be admitted to universities. However, the reform of boys' education in the 1860s provided an impetus for the placing of more emphasis on science and mathematics: in effect, Davies was proposing for women the traditional system that was already being replaced for men (Phillips 1990, pp. 251–53).

EDUCATIONAL REFORM IN FRANCE

In France in the late 19ᵗʰ century, several steps were taken to make education universal. One of the factors influencing the change was the prediction that obligatory attendance at school would bring about a significant reduction in the use of child labour in factories and workshops. Primary school began to be offered freely to everyone in 1874, and in 1880 manual apprenticeships were instituted. Children could also obtain instruction through private tutoring; evening courses offered by monastic orders; and manufacturing schools, such as the one run by the steelmakers Schneider et Cie at Le Creusot or the weaving school established in Lyon in 1886. Nevertheless, 90 percent of workers' children did not have access to education during this period, while the bourgeoisie monopolized access to the *lycées* founded by Napoleon I, which in turn provided the only route into universities.

Against this background, the passage of the Loi Camille Sée in 1880 led to "the most comprehensive state-controlled system of secondary education for girls that Europe had yet seen" (Albisetti 2004, p. 143). Camille Sée, the man for whom the law was named, had made an extensive study of girls' schooling in the United States, Switzerland, Sweden, Germany, and the Netherlands. He was sceptical that girls could master the same level of mathematics that boys did in the *lycées*, so he recommended that study in the new *lycées* for girls should centre on foreign languages. He also recommended that women be appointed as heads of these girls' *lycées*, despite his own findings that men headed girls' schools in other countries, and that coeducational schools had been successful, primarily in both the United States and the Netherlands (Albisetti 2004, p. 146), but also, for example, in the Italian city of Florence, where two girls aged twelve and thirteen, admitted to the Ginnasio Galileo in 1879, held their own academically and even had a positive influence on the boys, as models of behaviour to be emulated (Albisetti 2004,

p. 148). Another surprising omission from Sée's reform was the experience of the publicly supported, female-led, secular secondary schools for girls created in Belgium by the educationist Isabelle Gatti de Gamond (1839–1905). Albisetti (2004, p. 147) suggests that Sée may have been uncomfortable with what he felt was encouragement of free thinking among Belgian girls. The Loi Camille Sée also made no provision for boarding facilities for girls: the schools were to be for day attendance only.

In 1881, the École Normale Supérieure was established in Sèvres to train women teachers, but at the same time the *lycées* for girls were required to offer five-year programmes, instead of the six years that Sée had wanted so that girls could prepare for the *baccalauréat*, the entrance examination for universities. By 1905, only 14,777 girls were enrolled in the *lycées*, far short of the enrolments that had been envisaged (Albisetti 2004, p. 149).

In 1900 Anna Tolman Smith, an employee of the US Bureau of Education, compared the curriculum of the French *lycées* for girls with that of their US counterparts, and concluded that the former offered only "a type of liberal education essentially feminine," emphasizing, in addition to modern languages, domestic economy and hygiene. She suggested that training in modern languages and history performed much the same function as the study of the classics did, and praised the French schools "for providing training for marriage and motherhood" (Albisetti 2004, p. 151). In 1910, another American educationist, Frederic Farrington, reported on the year he spent visiting French schools. He noted that the enrolment of boys was three times higher than that of girls, and that expenditure on boys was seven times higher. He also remarked that the mathematics programme for girls was not "very extensive" and did not prepare them well for employment. Käthe Schirmacher, a feminist and educator from Germany—where all the states admitted women to universities—also spent time in France studying

the education system, and concluded that, since the girls'
lycées did not prepare them for university, they would have to
study privately if they hoped to gain access to higher education
(Albisetti 2004, pp. 155–56). In short, the Loi Camille Sée,
whatever its intentions, did not eradicate the backwardness of
French education for girls. Instead, French schools continued
to emphasise the preparation of girls for their traditional roles
as wives and mothers, in contrast to developments in girls'
education in the United Kingdom, Germany, the United States,
Canada, and a number of other countries.

EDUCATION OF WOMEN
IN THE UNITED STATES

The first coeducational college in the United States was
Oberlin College, which admitted women in 1837 and
awarding its first degrees to women in 1841 (Murray 2001,
p. 2). Others soon followed: Antioch College, in 1852; the
Normal School at the University of Wisconsin, in 1860;
the rest of the University of Wisconsin, in 1866; Boston
University, in 1869, the University of Michigan, in 1870; Sage
College at Cornell University, in 1874; and the University of
Chicago, in 1890. The number of women in coeducational
colleges and universities increased from 3,044 in 1875 to
19,959 in 1900. Many of these were state universities. A
number of single-sex women's colleges were also founded:
Vassar, in 1865; Wellesley and Smith, in 1875; Bryn Mawr,
in 1884; Mount Holyoke, in 1887; Barnard in 1889; and
Radcliffe, as the Harvard Annex, in 1894 (Albisetti 2000,
p. 476, and Murray 2001, p. 2). Clearly, the United States
and (as discussed below) Canada made much more rapid
progress on the question of equity in education than European
countries did.

Visitors from the United Kingdom were duly impressed
by what they found in the United States. Sophia Jex-Blake
(1840–1912), who visited Oberlin and Antioch Colleges in

Ohio, Hillsdale College in Michigan, and some schools in Boston and St. Louis, reported positively on the effects of coeducation and also observed that the mixing of students from different social classes stood in stark contrast to the norm in the English educational system (see Jex-Blake 1867). In spite of her observations, she expressed anxiety about the "deleterious effects of strenuous study on young girls" (Albisetti 2000, p. 479), and yet it was her experiences in the United States that inspired her to follow Elizabeth Garrett's example and become one of the first women doctors in England. The Reverend James Fraser, who reported on behalf of the Taunton Commission at about the same time as Jex-Blake, was surprised that both boys and girls studied classics and mathematics to high levels, although he continued to question US theory and practice on the education of girls.

In 1893, a British charity, the Gilchrist Educational Trust, sponsored a mission by five women to study the education of women in the United States (see Albisetti 2000, pp. 481–82). Their reports appeared in 1894, before the Bryce Commission finished its work. Mary Page investigated elementary schools, and her conclusions were fairly positive on the benefits of having girls and boys study together. On the other hand, Alice Zimmers, who reported on pedagogical methods, was ambivalent on the topic, citing one informant as to the economic gains that coeducational schools provided and arguing, cautiously, that the good effects outweighed the bad. Amy Bramwell and H. Millicent Hughes looked at teacher preparation and noted that most normal schools (teacher-training schools) in the United States were mixed, with more women than men attending. Finally, Sara Burstall, whose topic was the education of girls, commented on the ease with which male and female teachers worked together and suggested that "coeducation made discipline easier."

After another visit to the United States, in 1908, Sara Burstall was even more enthusiastic than before (see Burstall 1909). She remarked on how highly education was valued in the United States, presumably because of the democratic traditions of the country, the desire to create homogeneity in a nation of immigrants, and the attempt to moderate the effects of materialism. She also noted that the generous funding of schools led to free public education, as well as the provision of free textbooks to students, which were passed from one class to another year after year. Her report included statistics of attendance, indicating that 90 percent of the population did not attend school beyond the elementary level; that about 2 percent attended grammar or intermediate schools (for children aged ten to fourteen), though many students had to leave school early to start work; and that less than 1 percent of the population attended high school, which was then for students aged fourteen to sixteen. Of the 367,003 students who were attending public and private high schools in the United States in 1889–90, just under 50 percent (150,000) were female, and among these young women 21,000 were preparing for university entrance, a substantial number compared to other countries.

One more visitor who confirmed the success of coeducation in US schools was James, Viscount Bryce, who chaired the commission that is named after him. He commented revealingly that "the young men sometimes find the competition of the girls rather severe, and call them 'study machines,' observing that they are more eager, and less addicted to sports or to mere lounging" (Albisetti 2000, p. 481). This observation seems pertinent even now, when girls regularly surpass boys in academic performance in most subjects and more boys than girls tend to drop out of high school.

For all the progress being made in the United States, especially as compared to European countries, there were also indications of a reaction against increased opportunities for women. For example, the Universities of Wisconsin and Chicago, among others, decided to segregate their student bodies by gender and to reintroduce separate programmes for women.

Charles R. Van Hise, President of the University of Wisconsin from 1903 to 1918, was widely regarded as a progressive because of his support for extending university education to all classes, yet he publicly opposed the rapid increase in the numbers of women attending university, arguing that women's preference for certain courses would cause men to give them up; that male students themselves objected to the presence of women; and that women needed to be educated differently because their roles in society after they graduated would be different.

Similarly, in 1902, William Rainey Harper, the first president of the University of Chicago, enforced the segregation of women students, in spite of the opposition of a majority among the faculty. He also ignored the contradictions in his own views—between believing that women were the weaker sex and admitting that they were taking too many prizes—and in his own behaviour, since he supported the admission of women to graduate studies and appointed a few women to faculty positions (see Byers 1999).

EDUCATION OF WOMEN IN CANADA

As Paul Axelrod explains in his study of education in Canada in the 19th century (1997, pp. 1 and 10), initially, while "schooling mattered to many British North American colonists, . . . sheer survival mattered more." In a land dominated "by hostile weather, primitive modes of transportation, the relentless demands of farming or fishing, deadly diseases, and the eruption of war," children worked with their parents on the farm and in the household, and those parents who themselves could read and write taught their children at home whenever work was less demanding, mainly during the harsh winters. The main purposes for which a small minority of parents sent their children to school were to provide them with moral training and, in the case of boys, to secure a classical education for them, preparing them for practising law or medicine, or for being appointed to positions in the civil service. Girls were much less likely than boys to have access to education, and if they did, it was often through private tutors or in girls' academies, where, as in their English counterparts, they acquired the skills and knowledge considered appropriate for their gender and class.

The first coeducational school in Canada was the Grantham Academy, opened in 1829 in St. Catherine's (now in Ontario). In the same year, Upper Canada College was founded as a school exclusively for boys on the model of the "public schools" (in fact, elite private schools) of England. Education became increasingly widely available in both private and public institutions, for both the francophone and the anglophone populations, but for girls, in almost all cases, the "accomplishments" were still at the centre of the curriculum, including drawing, painting, dancing, sewing, and limited study of modern languages, but excluding Latin, Greek, and the sciences.

A prime example of the approach to the education of Canadian girls in this period appears in an address delivered in 1865 by a Mrs. Holiwell, head of the Elm House School for the Education of Young Ladies in Toronto, to the parents of her charges (reprinted in Prentice and Houston 1975, p. 250). She suggests that the goals in educating girls vary according to the ability and aspirations of their parents, whether to render a girl useful, ornamental, or intellectual. Since, in her view, all three qualities are necessary for young ladies, she proposes that the curriculum should include familiarity with history and geography (ancient and modern), arithmetic, and some knowledge of science and literature, the theory and practice of music, drawing or painting, modern languages such as French, Italian, or German, dancing, needlework, and "moral training." According to Mrs. Holiwell, "a perfect control of temper, a consideration for the feelings of others, respect for age and virtue, a modest estimate of self—these are the attributes of the true lady, and must be taught from infancy."

In the same year, George Paxton Young, a schools inspector and a Presbyterian minister who had moved to Canada from Scotland in 1848, was vociferous in his opposition to the idea that girls might also attend grammar schools, the elite secondary schools that prepared students for higher education. In his report "The Official Objection to Girls in Grammar Schools" (reprinted in Prentice and Houston 1975, p. 253), he wrote,

> I have been frequently asked whether I considered it desirable that Girls should study Latin in the Grammar Schools. It is, in my opinion, most undesirable. . . . I do not doubt the capacity of Girls to learn Latin and Greek; nor do I doubt that if they did learn these languages, the exercises would be beneficial. But I am not sure that, for the proper development of their minds, a different Course of Study might not be preferable. . . . classical study, as pursued in our Grammar Schools, is of no advantage to Girls whatever.

In spite of the objections of Young and others, girls did in fact achieve access to the same education as boys received in some of the grammar schools, and could even obtain university or college degrees. Dalhousie College in Halifax admitted women on the same footing as men from 1877 onward, and the University of New Brunswick in Fredericton also began to admit women in 1887. St. John, New Brunswick, already had a Young Ladies' High School (still in existence today as St. Vincent High School), which provided sufficient education to its students to prepare them for university studies (Prentice and Houston 1975, pp. 258–60).

AVAILABILITY OF BOOKS ON SCIENCE FOR EVERYONE

As Barbara Gates and Ann Shteir (1997, p. 8) have shown, "educated women interested in science still formed large portions of the audience at public lectures and read whatever was available to them." Books on science, many of them written by women for a general audience, were even more numerous in the 19th century than they had been in the previous eras. Writers continued to employ the style of letters, dialogues, or conversations, familiar formats for teaching science to young readers and to women of all ages.

In England, Jane Marcet (1769–1858) is said to have written the most influential and popular textbooks of the early years of the century. Her *Conversation on Chemistry, Intended More Especially for the Female Sex*, was first published in two volumes, anonymously, in 1805, and the author's identity was not revealed until after the 13th edition appeared in 1837. There were sixteen editions by 1853, in addition to three more books by Marcet on vegetable physiology, natural philosophy, and political economy (Phillips 1990, p. 110). Little is known of Marcet's rival Margaret Bryan (see Ogilvie and Harvey 2000, p. 298), except that she wrote extensively between 1780 and 1815, ran a school for girls, and taught them science to a relatively high level, using her own books, *A Compendious System of Astronomy* (1797) and *Lectures on Natural Philosophy* (1806), which she had published with the aid of Charles Hutton, editor of the *Lady's Diary*. The science classes in Mrs. Bryan's school concentrated on physics, mechanics, and chemistry. She did not claim that science would have liberating effects on the condition of women, and she avoided the secular approach of several other women writers. Other prolific authors were Rosina M. Zornlin, who published a number of popular textbooks on geology, geography, and travel during the 1850s, and Alice Bodington, whose efforts were focused on providing clear explanations of evolution for readers at the beginning of the 20th century.

An especially well-known woman writer of the period was Mary Somerville (1780–1872). Born Mary Fairfax, the daughter of a Scottish admiral, she was discouraged from learning mathematics by her parents, but she acquired books through her brother and studied in secret. In 1804, she married a cousin, Captain Samuel Greig, who was the consul of the Russian Empire in London. Her husband did not support her studies, but when he died three years after the marriage, he left her with a small pension, so she returned to Scotland with her two children and took up her studies once again. She later married another cousin, Dr. William Somerville, who encouraged her explorations of science and mathematics. Mary was fortunate also to meet and be tutored by Charles Babbage. Her scientific work began with experiments on magnetism in 1825, and the following year her first paper was communicated to the Royal Society by her second husband and published in its *Philosophical Transactions*. One of her main contributions was to explain the new French mathematical approaches in ways that could be understood in England, where they were little known. Her four principal books are *The Mechanism of the Heavens* (1831), an adaptation of Pierre-Simon Laplace's *Mécanique céleste*; *On the Connexion of the Physical Sciences* (1834), which she defines as a "group of sciences treating of matter and energy"; *Physical Geography* (1848), which was used widely in schools and universities for the next fifty years; and *On Molecular and Microscopic Science* (1868). In 1834, she was made an honorary member of both the Royal Irish Academy and the Société de physique et d'histoire naturelle in Geneva. She was the first woman elected to the Royal Astronomical Society in 1835, and received many other honours, including, seven years after her death, the naming of a women's college in the University of Oxford (Somerville Hall, renamed Somerville College in 1894, and coeducational since 1992). Mary Somerville was a staunch supporter of women's education and the first to sign a petition launched by the philosopher John Stuart Mill demanding suffrage for women.

ADA LOVELACE, JEANNE VILLEPREUX-POWER, AND SOPHIA KOVALEVSKAIA

In the 19ᵗʰ century, we continue to find outstanding women scientists and mathematicians. Three such women are profiled in this section, one from France, one from England, and the third from Russia. All three achieved substantial recognition in their day.

In France, Jeanne Villepreux-Power (1794–1871) is known for her work in marine biology (see Arnal, 2006). Starting as a dressmaker, she met and married an English merchant, James Power, in 1818. While living in Sicily, she taught herself all she could about marine biology, pioneering the use of aquariums for experiments. She eventually became a member of eighteen learned societies across Europe, and wrote several books. A crater on the planet Venus has been named Villepreux-Power in her honour.

In England in the 1830s, Mary Somerville encouraged Annabella Byron, the wife of Lord Byron, to provide a mathematical education for her daughter Ada, as the girl had obvious interest and capabilities in the subject. Mary introduced Ada to her former tutor Charles Babbage, a step that was to be central to Ada's career, and a lifelong friendship developed between the two women (Patterson 1987, pp. 208–14). Augusta Ada Byron, better known as Ada Lovelace (1815–1852), was born in London. Lord Byron deserted mother and child shortly after his daughter's birth and died only nine years later. Ada's mother encouraged her in her study of mathematics under a series of governesses and tutors. In 1835, at the age of nineteen, Ada married William, Lord King, who became Lord Lovelace three years after the marriage. During the following three years, Ada had three children, and this weakened her already frail physique. In 1840, her husband was elected to the Royal Society, providing Ada with access not only to its large library but also to fellow mathematicians. Working with Charles Babbage, she began to devise a calculating machine that would use cards to do basic arithmetic functions, and even solve trigonometric and algebraic equations. Her formulation of techniques, known to computer programmers today as looping, recursion, and selection, signalled the beginnings of computer science (Rappaport 1987, pp. 135–36), but her work was cut short when she died of cancer at the age of thirty-six. On Ada's innovative and visionary work, Dr. Betty Toole writes,

> Lady Lovelace's prescient comments included her predictions that such a machine might be used to compose complex music, to produce graphics, and would be used for both practical and scientific use. . . . Ada suggested to Babbage writing a plan for how the engine might calculate Bernoulli numbers. This plan is now regarded as the first "computer program."

In Russia, Sophia Vasilevna Kovalevskaia (1850–1891), daughter of an artillery officer and his wife, who came from a family of scholars that had emigrated from Germany in the previous century, was educated at home and, like Ada Lovelace, showed an aptitude for mathematics and science from an early age. In 1869, she went through a fictitious marriage that enabled her to leave Russia and study in Heidelberg in Germany, although she had to get permission from each professor in turn to attend classes at the university there. In 1871, she moved to Berlin to work with the mathematician Karl Weierstrass, but found herself entirely excluded from the courses at the university. By 1874, three years after her arrival in the German capital, she had written three doctoral dissertations. With support from Weierstrass and other mentors, she was finally granted a doctoral degree by the University of Göttingen in 1874, becoming the first woman to receive a PhD in mathematics. She wanted to return home, to Russia, but neither she nor her husband could obtain a post at any university there. In 1884, she was finally offered a post at the University of Stockholm in Sweden. According to A. H. Koblitz's brief biography of her (1987, p. 107), her main contributions were her proof of the theorem in partial differential equations, now referred to as the Cauchy–Kovalevskaia Theorem, and her work on the revolution of a solid body about a fixed point. In 1888, she received the Prix Bordin from the Académie des Sciences in Paris for this latter study, but only three years later, at age forty-one, she died of pneumonia, cutting short a brilliant and promising mathematical career.

PART III

—⚬⚬⚬—

Education and Careers in Science and Engineering

CHAPTER 8

Women in Engineering, Mathematics, and Science in the 20th Century

In the 20th century, universities were opened up to women in most countries, but their enrolment in science and mathematics courses did not progress on a continuum. Rather, it increased during certain periods and held steady or decreased at other times. The situation has, of course, become more complex than in previous centuries, when women were almost entirely denied access to formal education and to scientific societies, the gatekeepers for the creation and dissemination of scientific knowledge since the middle of the 17th century. In contrast to that age-old arrangement, women could now take part in any field of their choice, at least in theory. Yet there were, and still are, relatively few women in such fields as physics or computer science, or, indeed, most of the engineering disciplines. When universities and colleges began to open to women, few of them chose these subjects, and, although, as we shall see in this chapter, the situation has improved overall, women still predominantly study to become teachers or nurses or to enter other fields that are still dominated by women today.

In this chapter, the focus is mainly on women's participation in engineering, mathematics, and science in the United States and Canada, since the data for these two countries are readily available (see also Appendix 2). Some information is, however, provided on developments in Europe.

WOMEN IN MATHEMATICS AND PHDS IN THE UNITED STATES

Between 1862 and 1919, for example, colleges and universities awarded PhDs in mathematics to 474 men and 61 women. The question arises: why was the proportion of women in this group so low (just over 11 percent)?

In her book *Women Becoming Mathematicians* (2001), Margaret Murray recounts the lives of thirty-seven women who received PhDs in mathematics in the United States during the 1940s and 1950s, a period of particular interest not only because the number of women receiving PhDs actually dropped, to almost half of the number in the 1920s and 1930s, recuperating only in the 1980s, but also because this drop seems counterintuitive in view of the effect that the waging of the Second World War had on mathematics: while the number and the prestige of mathematicians increased, along with government funding, the number of women entering and staying in the discipline decreased substantially. The life stories of the women who chose careers in mathematics at this time are all the more intriguing, and offer lessons and even role models for women today. As in past centuries, a common factor contributing to the success of many of these women was cooperation from fathers, teachers, or husbands. Women still faced a great deal of social prejudice as they attempted to move through the educational hierarchy, but a little support from the right people could go a long way. Other issues raised in Murray's book include the importance of positive school environments; the need for mentors in the disciplines women were entering, above and beyond the best efforts of family members; and the detrimental effects of cultural stereotyping.

Above all, as Margaret Murray points out, the women in her study of the 1940s and 1950s could not follow the traditional pattern for a mathematical career, in which a true mathematician discovered his (very rarely her) talent early, continued through school and college to the PhD level without a break, and did his (even more rarely her) best work before he reached the age of forty. Murray contrasts this with the career pattern of the women in her study, who had diverse interests and commitments, usually took relatively longer to progress from completing their undergraduate work to receiving their PhDs, and often did their best work when older, in addition to coping with family responsibilities, whether caring for children, managing a household, or supporting older relatives. The traditional, male pattern was simply not an option for these women.

Sadly, this is still true today for most women in mathematics. A half-century on from Murray's sample, women in the United States today account for about 25 percent of PhDs in mathematics, a level only slightly higher than in the 1920s and 1930s. Skill in mathematics obviously forms a good base for careers in science and engineering, yet the proportion of women taking higher degrees in these disciplines is much lower.

WOMEN IN ENGINEERING IN CANADA

In Canadian universities, faculties and schools of engineering were among the last to see progress on enrolment of women, following in the wake of other male-dominated faculties, such as law, medicine, and business studies, which all achieved something close to parity between men and women by the mid-1990s. In this respect, Canada is typical of western countries, where women continue to be underrepresented in engineering studies.

The pattern of enrolment in the Canadian case can be followed over the course of almost ninety years by examining the participation of women in engineering studies at the University of Toronto, which has been collecting data on the question since 1920. It should first be noted, however, that the university has long had the largest faculty of applied science (equivalent to engineering) in the country, and that the average enrolment of women there has generally been higher than the national average, meaning that the data from Toronto may make the situation look better than it actually is.

Between 1924 and 1939, the average proportion of women in engineering studies anywhere in Canada was just 0.15 percent. There were between one and six women enrolled in the subject at the University of Toronto in every year during this period, but only three women obtained a degree. The first of these was Elsie Gregory McGill, who graduated in mechanical engineering in 1927, and the other two obtained their degrees in 1935 and 1937, respectively. The number of men enrolled in engineering at the University of Toronto nearly doubled over the period, from 442 in 1925 to 832 in 1938, and 1,788 men obtained degrees. The Great Depression did not act as a deterrent to female enrolment, except between 1936 and 1938, and the same can be said of male enrolment.

The number of women studying engineering at the University of Toronto doubled and then tripled during the war, rising overall from twelve in 1939 to nineteen in 1943. Five women obtained degrees. Men's enrolment climbed from 949 in 1939 to 1,639 in 1945, and a total of 955 graduated. However, while acknowledging that some of the data are missing, we can observe that the end of the war led to a significant fall in both male and female enrolment. For women, government incentives to return to the home or to their former jobs may well have been an important factor.

The available data on enrolment across Canada indicate a significant rise in both male and female enrolment from the mid-1950s until the end of the 1960s. It should be noted that the average national enrolment of women in engineering was still only 0.23 percent in 1950, 0.30 percent in 1955, and 0.84 percent in 1963, and it reached 1.84 percent, still a very low level, in 1970. At the University of Toronto, as mentioned, the proportion of women studying engineering was slightly higher throughout these years.

The period from 1971 to 1989 witnessed a more dramatic change. Some of the contributing factors were more active recruitment by universities; the influence of the women's movement; changing attitudes towards working women, more specifically towards women working in traditionally "male" fields such as science and engineering; a positive policy climate, resulting in the adoption of laws and court rulings aimed at ending discrimination and achieving equity between women and men; and the continuing rise in general female enrolment at universities. There was also a decline in male enrolment in the mid-1980s, which raised the proportion of women in engineering studies. The statistics for the period show that 3.6 percent of Canadian engineering students were women in 1975, 7.9 percent in 1980, 10.8 percent in 1985, and 13 percent in 1989 (see Frize and Heap 2001).

The next period may be called "the post-massacre years": on December 6, 1989, a young man entered the École Polytechnique, an engineering school in Montreal, and killed fourteen women with a semi-automatic rifle. Before this tragic event, a new national programme had been announced, on May 12: the Northern Telecom/NSERC chair for Women in Engineering. I held this position at the University of New Brunswick between December 11, 1989, and June 30, 1997. The chair's mandate was to help to increase the participation of women in engineering studies and in the profession. The massacre and the continuing low enrolment of women in the subject led to increased public discussion on the place and role of women in engineering, and gave impetus to the development of multiple initiatives, the largest of which was the creation of a nineteen-member committee to examine remaining obstacles and to suggest strategies to improve women's presence in these fields (see Canadian Committee on Women in Engineering 1992). Another key intervention in Canada was the creation of five regional chairs for women in science and engineering in 1997. On the research side, projects examined the aims and motives of young women choosing or not choosing engineering (see Lupart and Cannon 2000); the impact of computer courses and games on girls (see Crombie and Armstrong 1999, and Crombie et al. 1999); and the influence of parents on girls' career choices (see Wood 1999).

In spite of these encouraging developments, women are still far from reaching a critical mass, meaning approximately 35 to 40 percent of the student population. In September 2005, an article in the *Globe and Mail* entitled "Where Jobs Are and Students Aren't" (see El Akkad 2005) reported the declining interest in Canada's computer and information science programmes, at a time when the overall demand for information technology professionals was growing, and asserted that the plummeting demand for these degrees was also evident in the United States. It cited the conclusion of the Computing Research Association (CRA), which represents more than two hundred university departments of computer science, computer engineering, and related fields in both countries, that there had been a surge in the number of students in the late 1990s coinciding, with the industry's "dotcom bubble," but that then new enrolments dropped by 23 percent in 2002 and by another 10 percent in 2004. Data on university enrolment released by Statistics Canada in October 2005 confirmed this downward trend. While enrolment had risen in almost every field of study in 2003, the new data revealed that female enrolment in mathematics, computer science, and information sciences had dropped by 6.1 percent between 2002 and 2003, compared to 2.2 percent for male enrolments (see Statistics Canada 2005).

Meanwhile, overall female undergraduate enrolment in engineering schools and faculties levelled off at around

20 percent between 1999 and 2002, and then began to drop, like female enrolment in computer science, in 2003. Even more troubling was the decline in the number of female students enrolled in the first years of these programmes. The average was just slightly over 18 percent in 2001 and 2002, a decrease of 3.25 percent from the two previous years. In 2005, first-year female enrolment in several universities fell dramatically.

This recent downturn is especially disturbing in view of the impressive overall expansion of female enrolment in Canadian universities in recent decades. By 2003, women accounted for 59 percent of all undergraduate registrations, the highest proportion ever. The recent decline in female engineering enrolment follows twenty years of significant numeric gains by women in other fields of study. Women represented less than 8 percent of full-time undergraduate engineering students in 1980, but they accounted for 14 percent in 1990 (see Canadian Council of Professional Engineers 2004). Two main arguments had been used to justify the interventions that encouraged this rise in enrolment: the need to achieve equity by the elimination of gender discrimination, and the need for a larger pool of scientists and engineers to help Canada to become more competitive in the "knowledge economy" (see Solar 1997 and Heap, 2003). The serious decline that began in 2002 was observed across the country, except at a few universities, where it was less pronounced. This became a concern, and strategies to reverse this trend began to emerge (discussed in Chapter 10).

It is important to note that the proportion of women awarded degrees varies according to the fluctuation in the numbers of men in these programmes, so that it is possible to see more women in absolute numbers and yet also see the proportion go down because the number of men completing degrees has risen. The opposite is also true. Enrolment levels also depend on the rates of retention of students already in the programmes. In Canada, it has been common to observe that women do well in engineering. The proportion of women on deans' honour lists of students showing relatively high academic performance is frequently higher than the proportion of men in relation to the female and male populations studying.

PATTERNS AND TRENDS IN ONTARIO AND QUEBEC

Another approach to understanding the enrolment of women in engineering studies in Canada is to compare the figures for the country's two largest provinces, Ontario and Quebec (see Frize and Heap 2001). In Ontario, the number of women in undergraduate engineering studies increased from zero in 1924 to twenty-eight in 1964, representing an average increase of just 0.6 women each year, while for men the increase was 111 each year, or 15 percent. It was only in 1950 that women started to be seen more regularly in these courses, in both provinces. Then, between 1965 and 1985, years in which many recruitment programmes were created, the average number of women added each year was 79, representing an increase of 281 percent each year, while the number of men rose by 431 each year, or 8 percent. More recently, between 1979 and 1998 the number of men studying engineering in Ontario decreased by 6.5 percent, from 15,791 to 14,759, while the number of women increased by 140 percent, from 1,668 to 4,020 women.

As for Quebec, the number of women studying engineering increased from zero in 1924 to seventy-eight in 1964, an average rise of two women each year, while for men the increase was ninety-one men each year, or 17 percent. Between 1965 and 1985, the increase was 55 women each year (71 percent), while men's participation rose by 253 men a year (6 percent). Finally, between 1979 and 1998, the number of men studying engineering in Quebec went up by 15 percent, from 8,335 to 9,590, while the number of women increased by 132 percent, from 847 to 1,961.

It is noteworthy that, because the number of men rose faster in Ontario than in Quebec, the proportion of women was similar in both provinces by 1985, with the absolute number of women in Ontario increasing more rapidly than in Quebec. However, the proportion of women in Quebec was 1 percent higher than in Ontario by that time.

Statistics gathered by the NSERC/Alcan Chair for Women in Science and Engineering for Quebec between 1999 and 2005 (see Sévigny and Deschênes 2007) provide interesting information (albeit based on the data available, which are always a few years out of date), as we can observe that the decrease in the numbers of female engineering students in that province was almost exactly mirrored by a similar increase in the numbers studying in the health sciences sector. The data indicate that women are still predominantly choosing life sciences over physical sciences and engineering. Between 1999 and 2005, the number of women entering university in Quebec generally increased from 18,707 to 19,678, representing an increase of 5.2 percent in six years, yet the number of women students in the sciences and engineering fell from 3,292 to 2,396, a decline of 27.2 percent, with most of the decline occurring between 2003 and 2005 and mirroring the increase in the numbers of women studying health sciences. This adds credence to the assumption that many women seek careers where they see themselves helping people, and that, unfortunately, the sciences and engineering are still not seen as satisfying this need.

Between 1999 and 2005, men in Quebec continued to enrol in large numbers in the sciences and engineering, although the proportion of male students choosing these fields diminished slightly, from 45.8 percent to 44.5 percent, after having reached a peak of 47.9 percent in 2002. The number of men entering the first year of these courses decreased from 44.9 percent in 1999 to 38.2 percent in 2005, so that the proportion of men was nearly back to the level of 1999. (See appendix 2 for tables.)

WOMEN AND HIGHER DEGREES IN ENGINEERING IN THE UNITED STATES AND CANADA

The proportion of women among those who receive master's degrees in engineering in the United States (see U. S. National Science Foundation 2004) has tended to be slightly higher than the proportion among those completing undergraduate degrees, at 11.4 percent in 1986, 17.1 percent in 1996, and 21.2 percent in 2001. However, the proportions were lower at the doctoral level: 6.7 percent in 1986, 12.4 percent in 1996, and 16.9 percent in 2001. In Canada, the statistics are similar to those in the United States (see Canadian Council of Professional Engineers 2006), with higher enrolment of women at the master's level than at the undergraduate level: 21.6 percent of total enrolment in 1997 and 24.6 percent in 2001. At the doctoral level, 8.8 percent of those who received

Figure 8.1 Natural Sciences and Engineering Research Council: EKOS Research Associates Inc., Evaluation of the University Faculty Awards Program, June 2006.

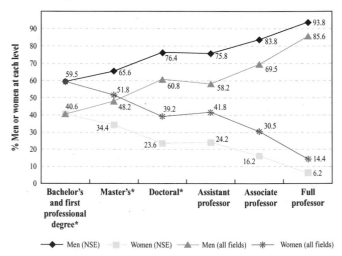

the degree in 1997 were women, while the number was 15.3 percent in 2001.

The number of women academics in engineering is not easy to assess. In both the United States and Canada, we find women at the lowest level, in teaching-only positions as instructors or on single-term assignments as sessional lecturers. Other women are adjunct professors, which means that they can do research at a university but do not have a tenure-track or tenured academic position, and no salary is attached. Still others have term contracts that may or may not be renewable. Among professors, whether assistant, associate, or full, women are, not surprisingly, predominantly found in the lower echelons. For example, there were fifty-one women with the rank of full professor in faculties and schools of engineering in Canada in 2001, out of a total of 1,376 (3.7 percent); eighty women were associate professors out of 678 (11.8 percent); seventy-nine were assistant professors out of 612 (12.9 percent); twelve were instructors out of ninety-two (13 percent); and thirty-seven were sessional lecturers out of 412 (8.9 percent).

Recent data show that not only in Canada and the United States but also in the European Union (see below), women leave academia at each level of progression from undergraduate degrees to full professorships. As a result, the numbers of men and of women who are qualified at the lowest echelon are similar, but the numbers of women fall off at each promotion level, while the numbers of men increase, in what has been called a "scissors" effect. (The Canadian data are presented in graphical form in Figure 8.1.)

WOMEN IN ENGINEERING IN EUROPE

Internationally, the availability of data on the enrolment of women and men in university engineering courses varies greatly, and there are also variations as to which disciplines are included in the definition of the sciences. For example, some countries include nursing and home economics, while other countries include only physics, chemistry, geology, and biology. Computer science is sometimes placed with engineering, but it can also be found as an independent Faculty, yet frequently included within the sciences. Some countries treat medicine and nursing as a single unit called health sciences, while others separate them. Another obstacle to understanding is that some countries still do not collect gender-based data. Caution is needed when attempting to interpret enrolment statistics, especially when making comparisons between countries, or even between different universities in one country.

As far as the European Union is concerned, data collected in 2000—when it had only twelve member states rather than the present twenty-seven—showed the same "scissors" pattern as in Canada: as the level of academic position rose from instructor to full professor, participation by women staff in engineering studies decreased, with the result that among full professors (or their equivalents in each of the twelve countries), more than 90 percent were men (see European Technology Assessment Network 2000).

In France, to take just one example from Europe, there have been several options for entry into engineering studies since the 1990s. One option is to study for two to three years to prepare for entrance into one of the so-called *grandes écoles*. Unlike the French national universities, which are obliged to accept all candidates from their given region who have passed the *baccalauréat* (university entrance examination), the *grandes écoles* rely on competitive written and oral examinations for which students must take appropriate preparatory classes. Another option is to study for two to three years at a university and then attempt the admission examination at one of the *grandes écoles*. A third option is study at one of the small number of specialized technical universities. Among students enrolled in engineering courses in the *grandes écoles* in 1984, 21.3 percent were women

(Charles 1991, p. 83), and the proportion has not significantly risen since then. The enrolment of women in the technical universities, meanwhile, was less than 5 percent of the total in 1972, 19 percent in 1991, 23 percent in 2000, and 25 percent in 2005. As of 2005, 15 percent of the 580,000 engineers in France were women (90,000 in all), and of those women 43 percent were less than thirty years old (see Association Française des Femmes Ingénieurs 2006).

As for broader comparisons, the Organization for Economic Cooperation and Development (OECD) reported in 2001 that the proportions of women among students receiving degrees in its relatively broad category of "engineering, manufacturing, and construction" were 32 percent in New Zealand, 31 percent in Slovakia, 30 percent in the Czech Republic, 28 percent in Hungary, Spain, and Sweden, 24 percent in France, Poland, and Turkey, and 19 percent in the United Kingdom. The lowest rates of participation by women were in Japan (10 percent), and the Netherlands and Switzerland (both at 12 percent). Since 2003, however, numbers have been falling in most of these member states of the OECD, just as they have been in the two member states we have focused on here, the United States and Canada.

LEVELS OF PARTICIPATION BY WOMEN IN SCIENCE

Information on the participation of women in the sciences is even more difficult to obtain than the data for engineering. In the United States, the National Science Foundation provides extensive statistics on degrees awarded in the physical sciences, mathematics and computing, earth, atmosphere, and ocean sciences, and biological and agricultural sciences, as well as psychology, social sciences, and ancillary disciplines. The physical sciences, which encompass both physics and chemistry, saw a large increase in the 1980s and 1990s, so that in 2001, 20 percent of university degrees in these disciplines were awarded to women. The proportion of women in biology and the earth sciences increased even more dramatically between 1966 and 2001. However, the reverse trend is being observed in mathematics, computer science, and even biology. In computer science, the proportion of graduates who were women decreased steadily from 40 percent in 1985 to just under 32 percent in 2001. The trends were similar in Canada and western Europe.

During the 1990s, girls performed at least as well as boys in all subjects in secondary schools in Europe, and in junior and senior high schools in the United States and Canada. In France, in particular, girls fared better at all levels, and more girls abandoned vocational courses in order to enter university education (Charles 1999, pp. 68–69). Furthermore, in Canada, the United States, and many countries in Europe, more women were enrolled in university than men (see European Technology Assessment Network 2000). In short, there is no issue of academic performance for girls when it comes to choosing science, engineering, or technology studies. Other factors must be at work.

Consider, for example, the situation in Canada in 2000, when women made up more than 40 percent of engineering students at the University of Guelph, but less than 30 percent at the University of Toronto and at Queen's University, even though all three are leading institutions and are located in the same province. The differential success in recruiting women into these traditionally "male" disciplines reflects how each university (and not just these three) can take steps to increase female participation. To maintain the gains made in recent years and continue working on narrowing the gender gap, above all in the sciences and engineering, we must assess what universities can do to make these disciplines more woman-friendly. The issues, and some possible solutions, are discussed in the next chapter.

⊶∞∞⊷

CHAPTER 9

Obstacles to the Entry of Young Women into Science and Engineering

This chapter presents a few of the many myths that still influence attitudes to women's and men's abilities, interests, and behaviours, as well as expectations about what careers are appropriate for each sex. It then addresses some of the other obstacles in the way of women's advancement.

As we have seen in previous chapters, throughout most of recorded history women have been channelled into domestic roles and barred from formal education, except, in some countries and some eras, at the elementary level. Women have also been excluded from the privileged circles that controlled the scientific enterprise, especially after science became more important in society, between the 17th century and the 20th century. Moreover, the voices of thinkers who were opposed to education for women and considered women as inferior beings were consistently louder than the voice of those thinkers who considered that women, if educated, could play the same roles as men.

The voices opposing women's education were amplified by the myths regarding gender roles that have been created over the ages and imprinted on the minds of most women and most men for many generations. It is not surprising, then, that gender stereotypes are still very much present in today's world. Myths are like legends: some people believe in them, but they do not represent reality. They exist because people feel comfortable believing in them, or prefer, with some complacency, to accept what the myth represents, especially if it fits their own views. It is realistic to say that it is easier to maintain beliefs in myths that support the status quo than to challenge long-held beliefs and change patterns of behaviour, in the hope of building a world of equality and opportunity for everyone. Debunking myths and legends can make people uncomfortable and uneasy, but clarification can help in making the world a better place.

MYTH 1: GENDER STEREOTYPES HAVE DISAPPEARED

It is interesting to observe, as we sit in an airport, a train station or a park, how parents treat girls and boys differently. Often, you can see a boy being given freedom and independence, and allowed to run around, while his sister is hugged or kissed and kept closer to their parents. Although not all parents act in the same way, this pattern of behaviour is still common enough to cause concern about the perpetuation of gender stereotypes.

In 1982, Alice Baumgartner-Papageorgiou published the results of a survey she had undertaken of two thousand young people, aged between eight and seventeen, in New Jersey. Her questions were ingenious: she asked girls what their lives would be like and what careers they would choose if they woke up the next day as boys, and she asked boys how things would be if they were to wake up as girls. Her summary of their answers is revealing and thought-provoking even today. The consensus among the girls was that, as boys, they would be "calm and cool, not allowed to express their true feelings, be rowdy, macho, smart-alecky, show-off more, and would be valued by their parents." They most frequently said that they would become professional athletes, construction workers, engineers, pilots, forest rangers, or sportscasters, and that "their lives would be better economically and status-wise, and they would enjoy more freedom, and have a better time, with less responsibility." It is sad that they did not think they could make these same choices as girls or as women. For their part, the boys in the survey were even more traditional in their thinking than the girls were. The boys said that, if they were girls, "they must be beautiful, know how to put make-up on, no one would be interested in their brain, and they were not sure if they would be appreciated by their parents." They expected that, as women, they would get jobs as secretaries, social workers, models, airline stewardesses, or prostitutes. Iin other words, they still saw women as being limited to roles serving others, and did not see them in scientific or engineering careers.

Baumgartner-Papageorgiou repeated the survey with a new set of boys and girls ten years later, in 1992, and obtained very similar results. Then, in 1994, Myra and David Sadker published the results of a similar study, in which they asked students from twenty-four upper elementary and junior high school classes in Maryland, Virginia, and Washington, DC, to write essays about waking up as members of the opposite sex. Among the many responses they received, one twelve-year-old girl wrote,

> When I grow up, I will be able to be almost anything I want, including Governor and President of the United States. . . . People will listen to what I have to say and will take me seriously. I will have a secretary to do things for me. I will make more money now that I am a boy.

Another girl wrote, "I would feel more on top. I guess that's what a lot of boys feel." The boys had different reactions.

Most found the thought of being a girl appalling, disgusting or humiliating, and completely unacceptable. One sixth-grade boy wrote, "If I were a girl, my friends would treat me like dirt." Some even mentioned committing suicide if they woke up as girls.

Clearly, attitudes and behaviours can and do change, at least to some extent, from one generation of girls and boys to another, and it is important to repeat such studies from time to time, in order to understand adolescents' culture and find ways to challenge these stereotypes. However, there is little reason to suppose that young people's attitudes have changed enormously in the fifteen years since this study was published, or that they are very different in other countries than they were then in the United States. From these and other studies, conducted over several years, we can infer that early socialization still imprints both boys and girls with some powerful stereotypes, and, in particular, that boys learn early to disrespect girls and women.

What is crucial in the context of this book is that these stereotypes do not cease to make their impact when boys and girls become men and women, but exert a powerful influence over educational and career choices. In 1993, the government of British Columbia published a report on aspects of gender equity in education and training, compiled on behalf of the Canadian federal government and the governments of five provinces (see Coulter 1993). In this report, Rebecca Coulter describes how sexist attitudes and behaviours operate both in homes and in schools, from birth to adolescence and beyond. Parents often display attitudes and expectations that differ for male and female children with respect to achievement in mathematics and science. Many sexist attitudes are so entrenched that they have become unconscious and unnoticed. This partly explains why many girls still underestimate their own abilities in mathematics and science, in spite of the fact that they currently perform as well as boys. Even in the first decade of the 21st century, it is still common to see girls dropping these subjects, especially physics, when they reach high school, because they do not see them as relevant to their own likely futures. On the other hand, boys are prone to overestimating their performance in these subjects and choose them in greater number in high school.

Recent research in Canada confirms Coulter's findings. Judy Lupart and Elizabeth Cannon (2000) have shown that most students decide by the seventh grade on careers that either include information technology or exclude it. Including information technology in a career was the first choice for boys, but for girls it was their sixth and last choice, and therefore unlikely to be selected at all. Similarly, in a longitudinal study of attendance at enriched mini-courses— one-week university courses offered to high school students in the spring—Moyra McDill and Shirley Mills (2002) looked at the choices that girls and boys made regarding engineering or science mini-courses and reported that

> age 13 appeared critical for girls with respect to choices made for the future . . . and little change is expected after that age. Teenage girls appear to have formed strong opinions by this point, not only with respect to general subject area, but also for specific topics within those areas.

It is obvious from school results that for girls, academic performance is not the main obstacle to studying mathematics, the physical sciences, or engineering. Instead, gender stereotyping, as it appears in the media, family life, school, and society, seems to be the main influence on the educational and career choices of both girls and boys. This gendered socialization has results similar to those of systemic discrimination. Some boys and girls end up studying subjects that seem appropriate for their gender according to the stereotypes, or because of parental or societal pressure, or through poor counselling. They might well have chosen quite differently, on the basis of their true interests and skills, if they had been given more information and encouragement. Clearly, there is a need to eliminate this sort of misdirection.

The scarcity of role models to inspire young women to enter engineering, computer science, and physics perpetuates the problem, but it is also unfortunate that some researchers are still obsessed with demonstrating the superiority of men, in general, over women. It is time to set aside these debates and focus on how we can encourage girls and boys, as individuals, to understand and respect each other and to consider many career choices, regardless of how these may still be viewed by some in contemporary society. For boys, this means encouraging them to consider teaching, nursing, or social work. For girls, it means encouraging them to consider careers in the physical sciences, computer science, or the various disciplines of engineering. If this was done more widely, and with a concern to get beyond stereotypes and address individuals' needs and interests, all students could reach their full potential and find careers that really are appropriate for them, instead of careers that seem appropriate because of their gender.

MYTH 2: BOYS ARE BETTER AT MATHEMATICS AND SCIENCE

The literature is rich in articles studying the widespread belief that boys are "naturally" better at mathematics and science than girls are. Most authors agree that before the age of thirteen girls and boys are very closely aligned in performance in mathematics, and, generally, in their overall academic performance.

Virginia Valian (1998, pp. 84–86) reports that gender differences in scores in international tests of mathematics and science were even smaller in 1992 than they had been in 1978. In mathematical computation there was no measurable difference, while in word-based mathematical problems boys in the United States did a little better than girls. However, this seems to have been a culturally based difference, since when scores were compared between boys and girls in kindergarten, the first grade, and the fifth grade in the United States, Taiwan, and Japan, the children in the United States were in third place in almost every one of the nine subcomponents of the mathematical tests, and never scored first. Japanese girls scored higher on all subcomponents than boys in the United States did. These results demonstrate that gender is not the main determining factor in mathematical performance. Motivation, schooling, and other environmental influences are more likely to have been responsible for the small differences, where they exist.

In a recent article on this question, under the bold title "Gender Similarities Characterize Math Performance," Janet S. Hyde and her co-authors (2008, p. 495) conclude that

> for grades 2 to 11, the general population no longer shows a gender difference in math skills, consistent with the gender similarities hypothesis. There is evidence of slightly greater male variability in scores, although the causes remain unexplained. Gender differences in math performance, even among high scorers, are insufficient to explain lopsided gender patterns in participation in some STEM [science, technology, engineering, and mathematics] fields.

The single factor that probably has the most negative impact is that, as mentioned above, girls tend to underestimate their own abilities in mathematics and science, while boys are prone to overestimating theirs. In November 1997, for example, the *Globe and Mail* reported that in Ontario girls were doing better than boys in the provincial examinations in mathematics, by at least four percentage points, in the third, sixth and eighth grades. However, while 58 percent of boys in the third grade thought that they were good at mathematics, only 46 percent of girls did, and in the eighth grade, while a similar proportion of boys still thought the same, only 35 percent of girls did.

Building self-esteem and confidence in girls will contribute to improving their actual performance and helping them to realize that mathematics, like science and engineering, may be appropriate for them. If, however, girls do not see these subjects as useful in their future career, or see them as "masculine," they may avoid taking them during high school years, unless the subjects are made obligatory. Pat Rogers (1988) has shown that adapting teaching methods to girls' interests, and their general preference for a cooperative rather than a competitive environment of learning, can lead to great results. Teaching these subjects with some link to girls' and, indeed, boys' real lives may make them seem far more relevant. For example, in explaining a mathematical concept such as an exponential curve to a high school class, the teacher could plot the minutes between contractions during the delivery of a baby. This might make mathematics more interesting to girls than just seeing equations and abstract concepts. Adding social relevance to the content of a course may also help to garner more interest from girls in subjects that tend to be perceived as "dry," and it is probable that many boys would also enjoy this approach. This is not to say, of course, that girls cannot do abstract work, but that this method may render the material more interesting and lively for them (Lafortune and Solar 2003, 83–84).

In general, young women at university, like girls at school, tend to be harder on themselves than young men or boys are. For example, at Harvard University (see Light 1990) it was found that women who abandoned their studies tended to have higher grade point averages than the men who dropped out. While women tended to have higher expectations of their performance, men tended to externalize the reasons for failure, blaming teachers, textbooks, or other factors. Women tended to internalize, feeling that they did not work hard enough or that they were not sufficiently competent in the subject. They often thought that they had failed even when their marks were fairly high, while lower marks for men do not get much reaction from them. At the University of New Brunswick, Marie MacBeth, a chemistry instructor who has since retired, collected data over a period of ten years on the performance of women and men in science and mathematics courses, measured by their grade point averages, and found that the women outperformed the men in all the classes in every year. She shared her findings with young women in undergraduate science classes in order to show them how good they could be and to enhance their self-confidence.

MYTH 3: SEXUAL HARASSMENT
HAS VERY LITTLE IMPACT

Dr. Ursula Franklin of the University of Toronto once suggested to me in conversation that young women in junior and senior high school probably need to invest 15 to 20 percent more energy than young men to get similar academic results. The extra effort is needed, Dr. Franklin suggested, in order to cope with sexism and harassment. This is easier to understand after reading June Larkin's book *Sexual Harassment: High School Girls Speak Out* (1994). Through interviews with teenage girls attending coeducational schools in Canada, Larkin, herself a former teacher, uncovered a widespread pattern of everyday sexual harassment of girls by boys, and found that little was being done to solve the problem. She asks her readers (Larkin 1994, pp. 13–14), "When so much of a female student's day is spent fending off diminishing comments, sexual innuendos, and physical pestering, how can she be expected to thrive at school?" and argues (pp. 22–23) that incidents of sexual harassment

> are logical products of a culture in which women are generally devalued, reviled and mistreated. . . . Despite the gains made by some women since the early 20th century, the

continual devaluing of women's work, the lack of women in positions of authority and decision-making, the continual resistance to women having control over their own bodies, the visual representation of women as sexual objects, and the disparaging jokes about blondes, mother-in-laws and bimbos are just some of the ways the diminishment of women remains embedded in our cultural attitudes and practices. Sexual harassers don't just hatch in high school; they have evolved from years of training in a society that conditions them to treat women as less important than men.

This may be one reason why girls are reported to do better in single-sex schools or classes than in coeducational settings. Single-sex schools or classes may also be a good thing for boys, as suggested in an experiment at Maynard Public School, an elementary school in Prescott, Ontario (see Laucius 2009). Boys and girls have been separated in the seventh grade, in order to help in closing the gender gap in boys' literacy and their scores on provincial tests. The school also seems to have stimulated girls' interest in science. Both girls and boys are said to have improved in all subjects.

The debate about coeducation in the 19th century (discussed in Chapter 7) was based on different arguments than the contemporary debate on the same question. Coeducation has been the norm for some time, in Europe as well as in the United States and Canada, and the curriculum has also been similar for girls and for boys, except earlier in the 20th century, when girls tended not to study Latin or take classes in woodwork or mechanics. The main argument now put forward in support of single-sex education is that boys and girls alike do better academically when they are separated into single-sex groups. Specifically, there is some evidence to show that boys do better at reading and girls do better in mathematics, technology, and science if they are segregated. It seems that in both cases the improvement occurs because the children no longer care what the opposite sex thinks of them or their performance, and so they can concentrate on the tasks at hand. Perhaps it is possible that good results can be obtained in coeducational schools, but considerable effort is required to make the curriculum gender inclusive, to build and maintain self-esteem and confidence among girls and some boys, to eliminate sexual harassment, and to ensure that each student can reach her or his full potential.

WOMEN AND COMPUTER TECHNOLOGY

There is also still considerable work to be done to break down the historically and socially constructed association of technology with masculinity, which, as feminist researchers in technology studies have demonstrated, has led to the dominance of white males in engineering and computer science (see Sorensen and Berg 1987, Sorensen 1992, and Wajcman 2004). The literature reveals the many dimensions and repercussions of this strong connection. Boys tend to learn how to use computers at home, while girls may have less access to them.

Moreover, young women tend to regard computers as tools rather than consider using them for leisure activities (Lafortune and Solar 2003, pp. 30–33). Computer games have been mostly designed by men, for boys and men. A study in the United States found that in 1997 there were around three thousand games aimed at boys as the intended market and very few intended for girls. According to Sheri Graner Ray, a computer game designer and director of a computer game company, game designers have been ignoring a potentially huge market. She discusses (see (Ray 2003) how gender approaches differ in game play preferences and provides suggestions for designers to create games that appeal to girls. She cites three main areas of gender variance. First, males respond most to visual stimuli, females to emotion and touch. Second, males like to tackle conflicts head to head, while females prefer compromise, diplomacy, negotiation, and manipulation. Third, males tend to be satisfied with visual rewards, but females require emotional resolution.

An interesting study of 145 fifth- and eighth-grade girls and boys (see Heeter et al. 2005) assessed whether games designed by a member of one gender were preferred by players of the same gender: the findings showed that the girls preferred games designed by girls, and the boys preferred games designed by boys. The girls reacted negatively to violent themes, and girls' games were perceived as better for learning. The girls preferred to solve puzzles, rather than exercise their eye–hand reflexes, while the boys preferred action content and fighter games, and some of the boys felt that such content is inappropriate for girls. Those girls who did like violent content preferred to see it in cartoon or fantasy forms, rather than in the realistic forms that the boys favoured. Carrie Heeter and her co-authors point out that their findings corroborate the evidence in other papers that discuss sex preferences in computer games.

Since girls at the high school and college levels generally have less "tinkering" experience than their male classmates, some of them express a higher level of stress and frustration when using computers. Boys are more confident in their computer abilities, whether this is justified or not, and consider girls to be less competent, an attitude that can act as a powerful deterrent to the development of self-confidence in girls (Lafortune and Solar 1993, p. 33). Researchers at the University of Ottawa (see Crombie et al. 1999) found that single-sex technology classes in high schools, using a curriculum that was more attractive to young women, filled up quickly, and that the results in the performance and confidence of the female students in single-sex classes were as good as those of male students in mixed classes. However, the female students in the mixed classes performed markedly less well than the other two groups.

Other studies show that men and women view computers differently, especially with respect to their aims and uses.

Men seem to be more interested in the process of using technology, whereas girls and women tend to be more sceptical of its efficacy. Many women prefer to use technology for communicating and for fostering interpersonal relationships. They seem to be reluctant to be part of the "geek" culture and do not wish to spend a lot of time in front of computers. Studies also confirm that unlike those men who think of computers as toys, many women share a broader view and privilege the links between technology and social context. If they choose to study engineering or computer science, it is with the purpose of using technology or computers to improve society (see Wharton 2001, and Margolis and Fisher 2002).

SOCIAL RELEVANCE AS A FACTOR IN WOMEN'S CAREER CHOICES

There is evidence that women look for a career where they can help people. Nursing is certainly a career with a visible connection to caring for people, and even now, years after the first male nurses were trained and appointed, in most western countries up to 90 percent of nurses are women. Women also represent more than 50 percent of new medical students in many countries.

If we could connect the problem-solving aspect of engineering to subjects in the life sciences, the proportion of women in engineering might well increase dramatically. An example suggesting that this approach works is found at the University of Guelph in Ontario, where courses in water resources, environmental engineering, and biological engineering have been gender balanced for years. The enrolment of women in the Engineering School at Guelph reached 43 percent in 1997, at a time when the university also had a food engineering programme with a high proportion of female students.

In a study of twenty women and twenty men engineers in the Vancouver area, Anne Van Beers (1996) found that women tend to choose a field that already contains a higher proportion of women. This could in part explain why chemical engineering has had gender-balanced enrolment for several years. Van Beers found that women's choices are influenced by a better knowledge of, or interest in, a particular field, and that women who are married also consider opportunities for their husbands when choosing a field. Some women said that they had a parent in the same field. One third of the women in Van Beers's study said that the choice of discipline was an expression of their personal values and social interests.

THE INFLUENCE OF SCHOOLS, HOMES, AND SELF-ESTEEM

High school guidance counsellors can play a critical role in the educational choices that boys and girls make. Unfortunately, some continue to discourage girls from seeking careers in engineering, because they still consider it to be a field for men (see Anderson 2002). Several women who chose to study engineering have told me that, even after they had decided to become engineers, their guidance counsellors tried to talk them out of it.

The influence and role of parents is not, of course, negligible. As Phillips and Zimmerman (1990) point out, "Mothers have lower academic expectations, and allow lower grades and achievement, for them [girls] compared to boys."

The myth of male superiority has limited access to education for girls and women and imposed strict gender roles that only the boldest and bravest of women were able to defy. In spite of the progress made in western countries over the past one hundred years—to the point where, in stark contrast to past eras, and to some non-western countries today, women now have access to secondary education and to university studies, and can hold public positions and enter the professions—several obstacles remain for girls before they can seriously consider a career in engineering, physics, computer science, or technology.

Boys are also hampered by social expectations, but in different ways. For them, the belief that being good at sports is more important than academic performance, and the fact that they generally overestimate their skills and talents, means that they tend to invest less effort than they should and so perform, on average, more poorly than girls. This explains in part why girls are getting better school results than boys and why more of them win scholarships to attend university. Different solutions are needed to help both girls *and* boys to reach their full potential.

CHAPTER 10

Recruitment and Outreach

A variety of successful strategies were developed in the 1980s and 1990s to attract more young women into the sciences, mathematics, and engineering. However, what works with one generation of teenagers may not be as effective with the next generation, so efforts to attract more women into these fields need to be evaluated regularly and fine-tuned to the needs and culture of each new generation. It is also important to ensure that women have positive experiences when they choose to enrol in these post-secondary programmes: success is more than just numbers.

RESPONDING TO DECLINING ENROLMENT OF WOMEN

Recalling that the enrolment of women in undergraduate engineering courses almost doubled between 1990 and 1995, from 12 percent of the total to 22 percent, we need to understand and address the issue of the recent decline in the numbers of women entering the first year of courses in almost all the engineering disciplines across the developed world. It is difficult to point to a single reason for this decline, and a combination of factors is most likely to explain the situation.

First, the serious decline in high-tech jobs after the "dotcom bubble" burst may well have affected the attitudes of many young people. Those whose parents lost their jobs would have found technology-related careers less attractive, while those whose parents kept their jobs would have seen them working many hours of overtime.

Second, there has been a resurgence of conservative values in society, which has been accompanied and reinforced by reductions in grants to women's organizations, the continuing display of gender stereotypes in the media, and a new preoccupation with the drop out rates of boys in high schools. While this is an important issue in itself, this preoccupation has tended to overshadow the problem of girls dropping out of science.

Third, as was discussed in Chapter 8, the decline in the enrolment of women in science and engineering courses has been mirrored almost exactly by increased enrolment of women in health-related studies. For women, this is not a bad choice, considering the ageing of the population and the increased need for physiotherapists, pharmacists, optometrists, and other workers in health-related fields. Women also see a direct link with helping people in these professions. Nevertheless, this trend represents a serious loss for science and engineering, a point that may not yet have been considered seriously enough by scientists and engineers themselves. Many advocates of recruitment efforts have been spurred on by the belief that a critical mass of women will bring enriching and complementary perspectives to designs, technological solutions, and the culture in these fields.

It is not difficult to take it for granted that brilliant women will make it in any field. However, stereotyping of women's abilities may have particularly acute repercussions for those women whose performance is average. Will they succeed to the same extent as average men?

OUTREACH ACTIVITIES IN NEW BRUNSWICK AND OTTAWA IN THE 1990s

Youth outreach activities developed in earlier decades now experience mixed success in attracting women into careers in science and engineering. A study of such activities in British Columbia, published in 1995, raised many questions about their impact and success (see Vickers, Ching, and Dean 1995). Its authors found that coeducational activities usually had more success in attracting boys than in attracting girls, and that the strongest positive impact on girls came from a single-sex programme called Girls in Science. Of almost 1,600 students surveyed, 44 percent of the young women indicated that medicine was their first career choice, as against 26 percent of the young men, while only 4 percent of the young women chose engineering as a preferred profession, as against 20 percent of the young men. Both the young women and the young men in this survey placed high values on "future job security" and "interesting work," but the young women rated "contribution to society" more highly than the young men did. Interestingly, the impact of parents was similar for both genders and was rated more highly than the impact of friends, teachers, guidance counsellors, or the media.

The majority of summer outreach programmes in Canada in the mid-1990s had unbalanced enrolments, attracting many more boys than girls. In 1993, I obtained a generous grant from the government of New Brunswick to develop a new approach, Worlds Unbound (WU), which I designed to attract equal numbers of girls and boys, from fifth grade to eighth grade, with activities that would pique the interest of both groups. Each student attended the day programme for a week (Monday to Friday), and six such weeks were held every summer. Two to three of the weeks were held in

French, depending on the numbers of francophone students taking part. The goals were to increase understanding of science and engineering; to improve the confidence and skills of students in these fields; to have a positive influence on the selection of courses in mathematics and science in high school; and to increase the likelihood of a choice of careers in these fields.

To ensure gender-balanced attendance, the policy was as follows. If the number of boys enrolling was greater than the number of girls, the extra boys were placed on a waiting list. When the parents of boys on the waiting list found a girl to join the programme, that boy was admitted preferentially over the others on the list. This allowed boys and their parents to be part of the solution, that is, to help with attracting girls to the programme. After the first year, the community knew that this programme was beneficial for both girls and boys, so it was rarely necessary to apply this policy: the proportion of girls attending settled at around 50 percent. However, the policy remained in place in case there was a downturn in applications by girls.

In the spring and summer of 1997, I carried out a survey to assess whether the summer camp had achieved its goals. Questionnaires were sent to all the students who had attended in the previous four years, and to their parents or guardians. In total, 162 sets of responses were received: forty-six from girls, thirty-five from boys, and eighty-one from parents or guardians. The overall results are presented below (Table 10.1). Not all the questions asked have been included, as some were designed to improve activities and were not related to the objectives of this book.

On the one hand, this survey obviously involved relatively small numbers of girls, boys, and their parents or guardians, in just one small Canadian province over a short period, so we must be cautious about interpreting its results. On the other hand, they do seem to support the view that a very positive impact can be made, on both girls and boys, when

Table 10.1 Summary of 162 sets of responses to a survey on the Worlds Unbound Programme, New Brunswick, 1993–97 (%)

Participants (46 girls and 35 boys)

1. How have your feelings changed towards Engineering and Science since attending Worlds Unbound's summer programme?

	Not changed	More positive	No answer
Girls	24	70	6
Boys	46	43	11

2. Has anyone tried to discourage you from doing math, science, or technology?

	Yes	No	No answer
Girls	2	87	11
Boys	6	91	3

3. Has anyone encouraged you to do math, science, or technology?

	Yes	No	No answer
Girls	80	7	13
Boys	74	23	3

4. Reasons given for the encouragement.

	Girls	Boys
I'm smart	9	14
Good field, many jobs	57	46
Fun, interesting	4	3
I am a girl	2	
I can make a difference	2	

5. What was your favourite activity at Worlds Unbound?

	Girls	Boys
Liked everything	4	11
Hands-on activities	42	40
Sports	7	3
Project (Goldberg machine)	20	17
Computers	2	14
Dissection	22	9

Table 10.1 (*continued*)

6. Did the experience affect your choice of courses in high school?

	Girls	Boys
Strongly	15	0
Some estent	37	43
Not at all	22	23

7. How has attending Worlds Unbound changed your ability to do math and science at school?

	Girls	Boys
Very positive	24	6
Mildly positive	20	31
Didn't change	33	34

8. What would you like to do when you grow up?

	Girls	Boys
Medicine	28	3
Engineering	7	14
Computers	4	11
Teaching	9	0
Law	4	6
Science	4	0
Biology	4	6
Architecture	2	6
Don't know	13	21
[Totals, science and engineering]	[19]	[31]

9. Did you enjoy the camp experience or were you bored?

	Very positive	Mildly positive	Bored	Missing
Girls	54	59	7	0
Boys	46	46	3	5

Parents/guardians (81)

1. Do you feel your child has a greater probability now of pursuing a science career than before attending the camp?

	Yes	No	No answer
Parents/guardians of girls	43	28	28
Parents/guardians of boys	49	26	26

Table 10.1 (*continued*)

2. Is one of the parents/guardians in the field of science, engineering, or technology?

	Yes	No
Parents/guardians of girls	41	52
Parents/guardians of boys	34	66

3. Does your child have access to . . .

	Girls	Boys
Computer	93	97
Internet	76	74
Calculator	89	100

4. Do you feel that the camp has had a positive impact on your child's math and science marks in school?

Parents/guardians of girls answering "Yes"   67
Parents/guardians of boys answering "Yes"   77

5. How has your child's attitude towards engineering, science, and technology changed after attending Worlds Unbound?

Parents/guardians of girls answering "More Positive"   72
Parents/guardians of boys answering "More Positive"   63

6. What was the most important result of your child's attendance at WU?

	Parents/ guardians of girls	Parents/ guardians of boys
Fun while learning	4	14
Opportunity to work with others	9	9
Built confidence, interest	43	40
Negative effect	0	6

7. Would you encourage your child to pursue a career in science and engineering?

	Yes	No	No answer
Parents/guardians of girls	70	9	21
Parents/guardians of boys	80	6	14

outreach activities are designed with a view to promoting equity in all aspects. Worlds Unbound succeeded in increasing the interest of the students in science and engineering, and many students reported increased levels of self-confidence in dealing with these subjects. Interestingly, the levels of preference for careers in medicine were lower than those reported for British Columbia in 1995 (see above), at 23 percent for girls and 3 percent for boys, but the gender gap was strikingly wide in both studies.

It is likely that before their summer experience, more girls than boys had had negative attitudes to, or lacked information about, these fields, so the camp experience may well have opened up new interests for the girls. For the boys, the experience seems to have reinforced their previous choices. Thus, a summer camp where girls and boys were present in equal numbers, and where gender equity was a pervasive policy, succeeded in making members of both genders, but girls in particular, more aware of careers in engineering and science.

Before I conducted the survey, I had received anecdotal evidence of the success of Worlds Unbound, with both girls and boys, through comments made by parents or guardians and by participants at the end of each week. However, the survey confirmed the extent to which each of the goals of the camp had been met. The gender balance among the participants and the instructors, and in leadership roles, and the careful selection of content with sensitivity about approaches that would work best with girls, all contributed to the success of the camp. In its first year of operation, everyone was attentive to the ways in which girls and boys reacted to visits to laboratories and to the level of "child friendliness" (or lack of it) in the presentations at those laboratories. Many improvements were made in the second year by removing a few relatively poor speakers and replacing them mostly with young women graduate students, who related well to the girls and boys taking part.

Thus, coeducational outreach activities can be made to work, although single-sex activities, if well-designed, can do even better. A half-day event called "Pinocchio's Nose" (see Frize et al. 1998) brought one hundred girls and their teachers from local schools to one of Nortel's corporate sites in Ottawa. The activities consisted of presentations, hands-on activities, a team game, and visits to some of Nortel's laboratories. This proved to be an extremely effective way to reach the girls and to open their minds to potential careers.

Large amounts of public and corporate funding have been invested in promoting science and engineering, so it is important to ensure that they reach equal numbers of girls and boys. All organizers of outreach activities need to evaluate their impact on both girls and boys, and to assess whether the participation of girls can be substantially increased. In particular, designing activities for girls only, until such time as obstacles and stereotypes disappear from society, and ensuring a mix of activities (hands-on components, career choice guidance, visiting people actually doing engineering and science, games), can be a recipe for outstanding success.

REASSESSMENT AND NEW OUTREACH ACTIVITIES IN OTTAWA

In response to the declining enrolment of women in engineering, it was important to reassess how to reach the new generation of teenage girls in junior high schools (seventh to ninth grades). Most universities and colleges have been conducting outreach activities for several decades, but what has been less common is evaluation of their impact on young women's career and educational choices.

In Ottawa, we redesigned the established Pathmakers Program and tested whether it was effectively communicating with a new generation of young women. We also wanted to see if we could make these young students ambassadors to other girls in their schools who did not have the opportunity to attend

the events. The goal was not to turn all these students into
engineers or scientists, but to erase the stereotypes attached
to these careers, open up more career choices for them, pique
their interest in research and innovation, and encourage them to
include science and advanced mathematics in their high school
curriculum, as this would open many career opportunities
to them, and not only in science and engineering. Another
objective was to develop their self-confidence in performing
engineering design activities.

The original Pathmakers Program was created in 1986
as a partnership between the University of Ottawa, Carleton
University, Algonquin College, and the four school boards in
the Ottawa–Carleton district. Female students in science and
engineering made presentations and did hands-on activities
in schools with groups of students ranging in age from eight
to seventeen, although more visits were done to classes
of teenagers. Most presentations were in real classrooms,
and usually to both girls and boys, with teachers present
and sometimes with guidance counsellors. The goal was
to encourage girls and young women to be informed about
and consider opportunities offered by engineering, science,
or technology careers. Another objective was to create and
sustain communication between role models and students in
elementary and secondary schools. The messages delivered
by Pathmakers were: stay in school; keep your options open
by enrolling in science, mathematics, and technology courses;
explore career opportunities; plan your future economic
independence; and become well-rounded, educated adults.
Pathmakers sparked students' interest through personal stories
and provoked excitement and wonder through demonstrations
and hands-on activities. Volunteers discussed gender
stereotypes and the appropriateness of careers for women.
They communicated well with young people of various ages
and learned how to speak about themselves, why they chose
their field, what courses and qualities are needed, and the
types of employment likely to be available in the future. They

shared their goals, discussed the rewards and obstacles, their likes and dislikes, and their future plans, and told the students that success could come through sufficient desire, determination and effort.

The results were dramatic. Enrolment of women in undergraduate engineering programs at Carleton University and the University of Ottawa doubled, from 12 percent to around 24 percent, between 1996 and 2000, particularly in the fields of electrical and mechanical engineering. Several benefits were identified for all the partners in the programme. Girls and young women met women who enjoyed non-traditional careers, learned about new opportunities, and worked on interesting design projects. Teachers gained new ideas, resources, and teaching methods, as presentations were designed to enrich the science curriculum. Pathmakers volunteers provided an inspiration and a catalyst for the success of the next generation, which gave the volunteers themselves a feeling of satisfaction. They also gained communication, interpersonal, and presentation skills useful for their future careers. Post-secondary institutions gained an increased diversity of students in their engineering courses.

A way had to be found to perpetuate outreach activities in schools by involving the schools themselves, both teachers and students, with some of our volunteers. The coeducational environment of the classroom was deemed to be less effective than single-sex events, especially when the topic was non-traditional for girls and young women. Pathmakers volunteers eventually graduated and left, in cycles of four years, and tended to be very busy students, so it was difficult to sustain the high level of activity reached in 2000, when there were more than eighty school presentations by 150 volunteers.

In response to the recent decline in enrolment mentioned above, a new initiative was created in Ontario by representatives from all the engineering programmes in the province, under the leadership of the Ontario NSERC/Hewlett Packard Chair for Women in Science and Engineering. The first meeting was held at Queen's University in Kingston in February 2005. The discussion led to the creation of an event called Go Eng Girl, a name coined by Dr. Moyra McDill at Carleton University for a mini-course on engineering for girls, You Go Eng Girl. The plan for the Go Eng Girl event was to hold it on the same Saturday in mid-October at all the universities in the province, in order to attract the attention of the media and reach a wide audience. Schools provided invitations to girls in seventh to tenth grades and asked them to bring a parent if they wished to attend. Girls and parents first heard dynamic speakers, both women engineers and students, and then the girls were involved in two hands-on design projects. The girls were divided into groups and accompanied by women engineering students, and their activities were geared to two age groups: students in the seventh and eighth grades, and those in the ninth and tenth grades. Meanwhile their parents heard speakers describing how to encourage their daughters to keep their options open by studying mathematics and science in their high school years. They also discussed how parents could make themselves aware of potential math anxiety problems, and consider the appropriateness of these careers for women. Parents also heard about the various scholarships available at the host universities. The half-day concluded with a lunch and a demonstration of designs by the girls.

In the Ottawa area, several other components were added to the provincial plan. Following our previous success with the Pinocchio's Nose events in the late 1990s, and the success of the Go Eng Girl event itself, we took into account the fact that members of the new generation are said to have shorter attention spans, want to have fun, and need audiovisual stimulation if their attention is to be retained. We redesigned the Pathmakers programme to encompass four main activities. We added an event called "Design Tomorrow's World," to which teachers in both urban and rural schools brought young women in the eighth to eleventh grades who might be interested in careers in science or engineering. The event was a half-day in a central location hosted, as before, by the University of Ottawa, Carleton University, Algonquin College, and the four school boards. Activities included making a movie using digital pictures and multimedia, designing a robotic arm, designing a water filtration system for developing countries, making an electromagnet, and identifying DNA. These activities were similar to those at the Go Eng Girl event, except that they involved teachers who might have some influence on the students' career choice. Moreover, the student participants were asked to repeat the activities with younger students when they returned to their schools. This provided peer mentoring between teenage girls. The teachers were very supportive, and the attendance went up from forty young women in the spring of 2006, to eighty in the spring of 2007, to more than 110 in the fall of the same year. In 2008, more than 220 young women and many teachers took part.

For both Go Eng Girl in Ottawa and then Design Tomorrow's World, questionnaires were issued to adult and student participants, and overall the responses were very positive for all the events. The girls and young women said that the activities were interesting and fun. They valued the exposure to a variety of engineering roles and were grateful that such an event had been organized for them. More than two thirds of them indicated that the event had increased their awareness and knowledge of career prospects in engineering, and their interest in the profession. Teachers confirmed that many of the students later enrolled in advanced mathematics and science courses. Several became ambassadors for Pathmakers, creating peer-to-peer encouragement and enthusiasm for engineering. These were some of the comments received from students:

> The presentations have increased my general interest in engineering. I was a person who used to believe engineering was all gears and bridges. I had fun and learned many new things.

It was really interesting to see what they do at [company name] as I've always seen the building but never really known that much about it. Overall it was very fun!

Very interesting, learned a lot. Overall I enjoyed the day and I will think more about science and technology as a career opportunity.

It was awesome! I loved the activity! Thank you.

The presentations are good and you should do this every year. Keep up the good work!

This has been a great experience for me. I am now seriously considering becoming an engineer, now that I have been here. I am grateful for this. Continue doing this and changing other girls' outlook on engineering.

It was a life time experience for me and I would like to thank everyone who was here to help us understand what is really going on around us in the world. It was awesome. You helped me pick my career. Thank you!

I have to say I wasn't very enthralled with any of the presentations. Not that they were terrible, but I think I would have preferred to spend my mornings as I usually do, learning about chemistry and art, and not coming to learn about engineering.

Even the few who were less interested or enthused at least gained knowledge about what engineers do, so that too has been a positive outcome. The teachers proved to be keen to continue with future classes. Many of these young students are years away from entrance into post-secondary education, so it will take some time to assess the impact on enrolment in engineering studies.

A third component in Ottawa consisted of industrial visits. Teachers and girls, and a few boys, were invited to visit corporations and national research laboratories involved in technological innovation. These visits were enjoyed by the students and by the teachers.

A fourth aspect was cooperative job placement of high school students. This provided a full term of five mornings or five afternoons of experience for the students. I have supervised eight such placements, with seven young women and one young man, in the past ten years. Several of these students later enrolled in engineering or science courses after graduating from high school, and several are now successful engineers.

CONCLUSION

Other out-of-school activities are important to pique the interest of young people. Shaunda Wood (1999) has described some effective ways for parents to encourage gifted girls (and boys) by bringing them to extracurricular science-based activities, providing books, assisting them with school work, travelling, and visiting parks and museums. In all outreach activities, it is important to profile women's achievements in science and engineering and to show how their contributions are adding enriching perspectives to research, design, teamwork, or the development of emerging fields. Recent literature also mentions how the participation of women in an enterprise has helped the bottom line for these companies.

Attracting increasing numbers of women into science and engineering is only one aspect of the problem. Retaining them and providing environments that allow the blending of a feminine perspective and approach with the existing culture and established norms of these fields also present major challenges. Some universities are very successful at retaining their women students, while others have dismal results. Several factors affect retention rates. In the next chapter, we examine the changes that could be made in order to improve them.

CHAPTER 11

Strategies to Attract
and Retain More Women

The focus of this chapter is on engineering, since it tends to have the smallest proportion of women among those studying it. However, most of the strategies that work for engineering would also work for computer science, physics, chemistry, or mathematics. I shall include some of my own observations, as a student in engineering in the 1960s, as a holder of a chair for Women in Science and Engineering in the 1990s, when I visited many engineering schools and faculties across Canada, and as an engineering professor for the past twenty years.

THE CULTURE OF ENGINEERING

G. J. Robinson and J. S. McIlwee (1991) point out that in order to understand how a culture functions, one must describe not only its values, norms, and styles of discourse but also the relations of power that underlie them. The question of power will become more visible when we discuss the engineering workplace, in Chapter 12. Here, we look at aspects of the culture, such as values and norms. The culture in an engineering faculty affects the environment in which the students learn and succeed (or fail), and is one of the factors that definitely need to be improved if these faculties are to attract and retain more women. There are other underrepresented groups in engineering, such as persons with disabilities, visible minorities or persons with different sexual orientations than the majority, but the focus in this book is on the gender component. If we succeed in improving the environment for women, it will likely be improved for everyone, even for the men who represent the majority in these disciplines. It is perhaps worth noting that none of the changes proposed in this chapter would endanger the accreditation by the profession of any engineering course.

Sally Hacker (1981) discusses the results of a study she carried out in a prestigious institute of technology in the United States through observations in classrooms, seminars, and social gatherings, and through a number of in-depth interviews with engineering faculty and with a comparable sample of faculty members in the humanities. She writes (p. 345),

> Engineering faculty ranked technical expertise as more valuable than knowledge of social relations. They described social sciences in womanly terms: soft, inaccurate, lacking in rigour, unpredictable, amorphous. Very few felt inadequate because they lacked knowledge of social relations or social systems. Almost all, however, felt engineers more qualified than most to move into management. They perceived little difference between managing people and managing technical systems.

The culture described by Hacker in 1981 seems very much like the one described by Robinson and McIlwee in 1991 (1991, p. 405):

> The culture of engineering consists of three components. First and most important, its ideology emphasizes the

centrality of technology and of engineers as producers of this technology. Second, it stresses acquisition of organizational power as the basis of engineering success. Finally, it requires that interest in technology and organizational power be "presented" in an appropriate form—a form closely tied to the male gender role.

Robinson and McIlwee's findings indicate that the culture of engineering changed little between the time of Hacker's study and theirs. They also note that the culture of engineering values behaviours and orientations consistent with the male gender role. They argue (p. 406) that

> engineering competence is a function of how well one presents an image of an aggressive, competitive, technically oriented person. . . . To be taken as an engineer is to look like an engineer, talk like an engineer, and act like an engineer. In most workplaces, this means looking, talking, and acting male. Of particular importance is to convey an image of hands-on competence. Few things are more closely tied to the male gender role than mechanical activities—using tools, tearing apart machinery, and building things. A fascination with, and desire to discuss at length, these activities is part of the culture of engineering's interactional display that works against women.

Two more scholars, K. H. Sorensen and A. J. Berg (1987), asked Norwegian engineering students to rate symbols (artefacts) that had some association with particular engineering disciplines as either masculine, neutral, or feminine. The results may explain the clustering of women in certain disciplines. Symbols rated as masculine, by both male and female students, included an excavator and a cement mixer, associated with some aspects of civil engineering, even though only a small proportion of civil engineers work directly in construction; a smelting furnace, linked to mining and metallurgical engineering; and an electrical drill, linked to mechanical engineering. Symbols rated as neuter included fibre optics, a microscope, a dentist's drill, and scales, each of which can be linked to chemical and biological engineering, fields where we generally find women enrolled in similar numbers to men. Symbols rated high on the femininity scale included a machine for packing fish, a typewriter, and a telephone exchange system, none of which has close links with engineering.

It is possible to infer from this study that women perceive symbols associated with chemical engineering as neuter or feminine, and enrolment is almost at par between men and women in this subdiscipline. On the other hand, it is not difficult to guess that symbols associated with mechanical engineering, civil engineering, and construction would be perceived as masculine. The associations made between these symbols and certain engineering disciplines can deter young women from choosing careers in these fields, considering that many of them are reluctant to be seen as masculine.

It is not surprising that many women will continue to avoid such careers until the image of the profession is more in tune with their interests and self-images. Demystifying the work of engineers, especially in fields where women are least represented, and demonstrating the human side of what engineers do may create images through which young women can envisage comfortable roles for themselves.

On the culture of engineering and technology, K. H. Sorensen (1992) has also written that "technology is decisively shaped by social relations, i.e. the political and economical needs of managers in industry, or by military interests." The studies cited here suggest that technology is characterized by masculinity and influenced by masculine interests. This is both a cause and an effect of male dominance in engineering and computer science, especially in the high-technology sector. As Sorensen writes,

Men's role has been as developers and principal users of high tech, whereas women's role has been more limited, as users, and mostly of low tech. Moreover, boys seem to be more fascinated by the process of using technology, whereas girls are more sceptical of its usefulness and many resist getting involved in its development.

OBSERVATIONS FROM THE 1960S TO THE 1990S

My upbringing was in a family where my father was a librarian and both my parents were published authors. My brothers and I met several writers and artists, friends of our parents, when they visited our family. When I was ten years old, I wanted to know how humans would get to the moon, and it was my passion for science, mathematics, and solving problems that enabled me to focus on my goals and persist in pursuing a degree in science and engineering. In order to survive in engineering, I isolated myself from the dominant culture and chose individuation rather than fitting in.

My experience of the culture in engineering, as the first female student in the Faculty of Applied Science at the University of Ottawa between 1963 and 1966, fits with the observations of Sally Hacker and the other scholars quoted above. During my student days, it was common to hear engineering students disparage students in arts or in commerce (the old name for what is now called business administration), assuming that engineering studies were too difficult for them. They also felt that anyone who failed in engineering could easily transfer to commerce or arts instead. I found that being the only female in a faculty dominated by self-confident men was not easy. It felt lonely, being different from the others in the class and frequently an object of curiosity. Some professors, deans, and peers were somewhat puzzled by my presence and did not seem to know how they should treat me, having never seen a woman in their classes until then.

However, they did do their best to be supportive and kind, and to make me feel comfortable. My own view was that I wanted to be treated just the same as everyone else, but that may not have been clear to everyone.

One interesting observation I made in the 1960s was that, when there were very few women around, I felt like a mascot, surrounded by "brothers," except for one of the students in my class, who was arrogant and mean. One day, I had to skip class because of a mild accident the day before, when I fell off a horse and was too sore to walk to class. Two students from my class came to my apartment to ask if I was abandoning my studies, and said that things were not the same without me in the class: they wanted me to return. I reassured them that I was not quitting and they offered to take notes for me during my absence. I was touched by their concern and was more determined than ever to graduate in electrical engineering.

On the other hand, the Lady Godiva Parade, for which a woman was hired to ride naked around campus on a horse, and the engineering song were very much part of the engineering culture in Canada in the 1960s, and even into the 1970s and 1980s. The last such parade appears to have been one at the University of British Columbia as recently as 1991, and Lady Godiva is still used as a symbol by some student groups on Canadian campuses even today, in complete disregard of her real historical role, and in ways that still demean women.

From 1970 onward, there were more women studying engineering, though not necessarily in every discipline, and I felt less lonely than I had as an undergraduate. However, there were few women professors in the classrooms. When the enrolment of women reached 15 percent, in around 1993, I could hear a few people saying that women were "taking over" and some backlash began to appear against incentives such as the Canada Scholarships offered by Industry Canada. Some men complained that it was not fair to give half of these scholarships to young women entering science and engineering, since there were so many more men entering these fields. The point they missed was that all these young women were deserving of these awards, and their numbers in science and engineering were increasing steadily, not only because of such incentives but also as a result of the more intensive recruitment activities that began in the 1980s. Another backlash was seen against the Women Faculty Awards introduced by the Natural Sciences and Engineering Research Council (NSERC), which provided partial salary and research grants to outstanding women who were offered faculty positions in science or engineering faculties in Canada. Both of these initiatives had received some positive responses, but the backlash endangered their existence, and they were cancelled in 1995. In 1997, Women Faculty Awards were reintroduced in a slightly different form, as University Faculty Awards available to Aboriginal candidates as well as to women. The NSERC's initiatives and the proactive recruitment of women faculty by some deans and department heads contributed to increasing the national average proportion of women professors in this field from 2.2 percent in 1991 to nearly four times that number in 2000. To compensate for the cancellation of the Canada Scholarships, several universities in Canada began to offer entrance scholarships to outstanding women students who applied to study engineering. Of course, these measures were meant to be temporary, until the number of women reached a critical mass.

One might have expected that things would improve at the beginning of the 1990s, but in my work as Chair for Women in Science and Engineering I heard many stories from women engineers and women students that suggested they had not really improved at all. Some of them said that professors had told them their place was at home, with a family, supporting a husband's career, and that they should not be wasting the professors' time and taking the place of male students. In the mid-1990s, a young woman in the second year of her undergraduate studies mentioned to me that she had overheard two young male faculty members in a corridor saying, "If we take on women graduate students, we should make them sign a promise that they will not have children." She asked me if that meant she should get herself sterilized before entering graduate school. She could not believe that such sentiments were still being expressed. In the same year, during a conference on women in engineering, one young assistant professor made a comment to another professor who repeated it to me later: "Why would a conference on women in engineering be needed? We don't need more women in engineering. We want our women at home, supporting our career." The man who said this had a stay-at-home wife and two young children.

When the proportion of women students reached 20 percent, several deans of engineering stated that nothing more could be done, since they had reached a "natural ceiling." My own view was that there was no "natural ceiling," just the subtle barriers created by the culture, the curriculum and teaching style, and the climate for female students. When changes occurred in these aspects of engineering studies and workplaces, then more women who had skills and interest in this field would enter it. Any ceiling there may be is constructed by engineers themselves, and in particular by the gatekeepers of the profession.

DIFFERENCES AMONG
THE SUBDISCIPLINES

Another aspect of the culture of engineering that is more delicate but should be discussed is whether certain subdisciplines in engineering are more highly valued or respected than others. Sally Hacker (1981, p. 345) writes,

> At the institute, I found that fields within engineering were also ranked, informally, along an "earthy–abstract" continuum. Electrical engineering (EE) carried more clout and status than, say, civil engineering. The former was

> considered "cleanest, hardest, most scientific"; the latter
> was far too involved in physical, social and political affairs.
> Most engineers agreed with the stereotype of EE, although
> those outside that field resented its power and status,
> merely because the field was closer to abstract science.

I was not surprised to read this. In the 1960s, I chose electrical engineering for my field of study not only because I liked mathematics but also because of the status attached to the subdiscipline. However, later, when my role as a chair for Women in Science and Engineering brought me into contact with people in other subdisciplines, such as mechanical, civil or chemical engineering, I learned to respect all these fields for what they could bring to humanity, and the previous status ranking of the various fields disappeared from my thinking.

However, there are still some engineers who think in the old way, labelling fields and topics as "hard" or "soft," with accompanying value judgments. I remember a male colleague in a Canadian university who heard that his department was considering a leading female researcher from a well-known university in the United States. He was opposed to the idea because, he said, her research was "fluffy." That adjective expresses the disdain with which faculty members like him still view research done by women. Her research was considered excellent by the university she had been working at, but it had no value in the eyes of a traditionalist engineering professor. Not surprisingly, no women had ever been hired by that department. The applicant from the university in the United States was offered the position, but she had heard what the professor had said, turned the offer down, and went elsewhere. This happened, not in the 1960s, but in the year 2000.

This aspect of the culture of engineering seems to be less common as newly emerging fields are becoming more established. I am sure that there are still people who think this way, but there are not so many among students as there used to be. Unfortunately, such traditionalists may still be found taking part in awarding grants or prizes, and the "old guard" still see the "hard-core," more abstract fields, such as electrical engineering, as more valuable than the newer multidisciplinary fields, such as biomedical or environmental engineering, which they probably consider as "soft," and more like social studies. It was certainly very difficult to get research grants in biomedical engineering in the early 1990s, but since the mid-1990s this has not been a problem. In fact, in many areas multidisciplinary research teams are encouraged by the granting agencies in Canada and the United States.

IMPROVING THE CLIMATE FOR WOMEN

The learning experiences of women (and of members of minority groups) still vary from class to class, and from university to university, but some common features recur. In many classrooms teachers make eye contact more frequently with male students than with female students, adopt a more attentive posture towards male students when they talk, often refer to them by their first names, call on them to answer more difficult questions, and use not only language that assumes that all students are male but also examples that reinforce negative stereotypes of women (see Martin 2000). Some professors may ignore female students' attempts to contribute in class, or may not intervene when some students demean others. This attitude on the part of professors, even if it is inadvertent or negligent, only adds to the problem. In the classroom, professors and instructors must do more than just show sensitivity to the diversity of student populations. As role models, they should be active participants in advancing the concepts of fairness, equity, and respect.

Matters are not necessarily easier or more pleasant outside classrooms. Demeaning rituals reminiscent of boot camps still occur, although they are less frequently reported, and sexist or demeaning jokes or comments are still made from time to time by students as well as by professors (see Ingram and Parker 2002). Occasionally, student publications carry material that is demeaning to women and inappropriate for institutions preparing students for entry into professions. It is especially distressing when women are involved as editors or authors of such material. This may seem to them to provide a way to demonstrate that they "belong," or they may be unaware of the demeaning aspects of what is published, having been desensitized by the culture around them.

Recently, however, more attention has been paid to the need to use gender-inclusive language in classes and at conferences. While the culture has certainly improved over the years, there are still instances of the "chilly climate", inside and outside the classroom, as described by Roberta M. Hall and Bernice R. Sandler (1982). In her thesis, as well as several articles, Karen L. Tonso (1997 and 2006) has discussed the sexism and violence she has observed in classes and in team work in an engineering school, and the silence about these occurrences, particularly from the women. She interviewed several male and female students and found that the men did not see the sexism or the violence, although some of them mentioned that it was natural for them as men to be on campus. Tonso writes (1997) of her own experience in engineering:

> When I entered engineering over 25 years ago, becoming an engineer seemed to imply that I could go out into the world and assume male duties. That presumption comes from an earlier day and age. Times have changed, and we must now understand that being forced to adopt, in an unexamined way, these culturally and historically taken-for-granted ways of construing professional identity is a violence against women that not only denies our existence, but also denigrates the importance and honour of women's work. To expect that women "check their identities at the door" constrains our participation—makes us silent and invisible.

She then goes on to provide advice about how teams should be formed. She recommends that a critical mass of women be chosen for each team, and because there are so few women in these classes, some teams will have only men. Tonso suggests that teams of three should comprise two women and one man, and teams of four should include at least two women. In a more recent article (Tonso 2006), she discusses the teaching of design and the organization of teams that would provide the best results. She recommends that instructors

> teach how to respect one's team-mates and the client's needs, organizing teams to reduce conflict, balance gender composition (or have all-men or all-women teams), and improve trust; incorporating peer feedback throughout the duration of the team, and using it to mediate disrespectful interactions and unacceptable practices; including a wide range of teamwork expectations in grading practices; and better training faculty about ways to incorporate teamwork.

In the 1990s, classes in engineering were generally quite large. Some of the classes I taught had more than 400 students, which made the teaching more impersonal and favoured a traditional pedagogical style. This created a more difficult climate for all students, but it could be particularly hard on women students, many of whom prefer more personal approaches. In engineering and the sciences, the teaching style is still mostly along the lines of traditional lectures, yet research has shown that other approaches, such as a cooperative learning style, can be more effective. This is not easy to apply to large groups, but there are ways to accomplish it, such as having the students work in groups to solve problems.

Long term structural and cultural changes are necessary to make schools and faculties of engineering more hospitable to female students. This represents a considerable challenge. It will obviously take a critical mass of women and considerable determination on the part of university administrators to create the policies and the plans to achieve a better gender balance. On the other hand, women may stay away unless the overwhelmingly masculine culture of engineering changes. As Ingram and Parker (2002) point out, engineering education has become a symbol of male culture, not only because of the pervasiveness of masculine values and interactional styles but also because of the male-dominated structure that supports it.

MULTIDISCIPLINARY APPROACHES

Many women enjoy multidisciplinary studies, especially if they see such studies as training them to do work that benefits people and society in general. Adding such subdisciplines as biomedical engineering alongside traditional courses in electrical or mechanical engineering, or environmental engineering with civil or chemical engineering, works well as a strategy for increasing the numbers of women enrolling. Yet curriculum content tends to be rigid; however multidisciplinary choices are increasingly being offered.

The Carnegie Mellon School of Computer Science, for example, which is ranked as one of the best in the world, has chosen to accommodate a wide range of computing experience among its incoming students, a decision that has proved successful in attracting female students (see Margolis and Fisher 2002). Carnegie Mellon also promotes interdisciplinary courses. At the University of British Columbia, the Computer Science Department has introduced a course on human–computer interaction that has had significant success among female students (see Hancock, Davies, and McGrenere 2004). These positive results suggest that women students tend to be more concerned with the societal context of technology than their male counterparts.

One major problem associated with engineering education is that theory is, in effect, seldom connected to real-life applications. Engineering education also values facts more than intuition and rationality more than emotional expression, although it is often intuition that leads to innovative ideas, and there is considerable literature on the benefits of using "emotional intelligence."

CURRICULUM REFORM

There is a broad consensus amongst researchers on the need to transform the curriculum and the prevailing pedagogical practices in engineering. A recent study of communication styles in engineering classrooms at the University of Manitoba (see Ingram and Parker 2002) concluded that instructors needed to develop sensitivity to achieving inclusiveness, which is essential if women are to participate actively in student teams. The study showed that it was the male-dominated teams that tend to bring out more gender-linked behaviour in both men and women.

The goal of making the curriculum friendlier to women has been discussed by several authors. Sue Rosser (1990, 1997, and 2004) has described several ways of creating a more "woman-friendly culture," arguing that these changes would also benefit male students. Sheila Tobias (1990), in a study of why some students, both male and female, abandon their undergraduate science courses, sometimes even midway to graduation, suggests that a massive restructuring of the science curriculum in elementary, junior high and senior high schools is needed to produce improved scientific literacy. She presents ethnographic notes on six graduate students and one professor, all in non-scientific fields, who sat in on college introductory classes in chemistry and physics to "explore their personal encounters with the courses and classroom culture" (Tobias 1990, p. 5). Her book also includes discussion of two studies, one of Radcliffe students who dropped out of science courses, the other of a cohort of students at the University of Michigan. Tobias concludes that those students who dropped out were bright, but they did not feel that they belonged in the culture they found in these classes. Some bright women, and men, stay away from science and engineering studies altogether. If more such students remained in these fields, the environment might become friendlier for everyone.

Tobias also suggests that class size affects teaching style. Large classes have long been taught by means of traditional lectures in large halls, and it is not easy to apply small-group self-study methods to large classes. She also recommends exit interviews with students who decide to drop out (Tobias 1992, p. 87). These would help some of them to reconsider and return, help others to find courses that better fit their interests, and also enable faculty to discover what can be done to improve the classroom climate and the quality of teaching.

Institutions are now responding to the findings of these and other authors. For example, the Association of Professional Engineers of Ontario (PEO) has produced a kit for gender-sensitivity training in engineering faculties and schools. The University of Ottawa Women in Engineering Research Group (UOWERG), led by Ruby Heap with researchers from social sciences, education, and engineering, is conducting an interdisciplinary research project to explore the links between efforts to make undergraduate engineering education more socially relevant and the retention of female students. This research will help in defining long-term improvements that can be made to increase the participation of women and improve their educational experiences.

There are simple ways to link theory and real-world applications in the classroom. For example, the teaching of electrical circuit theory, which requires students to use abstract mathematical equations to solve circuit problems, can be very dry, as can be seen in the textbooks on the topic. However, in the second-year course on electrical circuit theory that I taught for many years, things became lively, and students made connections with the real world, when I showed them how an RLC circuit, consisting of a resistor, an inductor, and a capacitor, can resuscitate a person who has had a cardiac arrest, since it is the basis of the design of defibrillators. I also presented other biomedical applications. Several students told me that they had understood the material better once we had discussed these applications, and they also found the course more interesting. Several of them chose the biomedical field for their final-year projects, and some even entered graduate studies in this field. I suggest that every course in science, mathematics, computer science, and engineering can be modified slightly to help students to connect their studies with social realities without losing any of the theory that still, of course, has to be transmitted to them.

Attracting more women into graduate studies and hiring more women faculty would also have a positive impact. Increasing the number of women faculty is linked to the success of attracting qualified women into graduate studies (as will be discussed in Chapter 12).

SUPPORT ACTIVITIES

Support for the retention of female students can be provided in several forms. Appropriate counselling or advice on course choices can help to ensure the students will have a good fit between their studies and their interests and abilities. If women enrol in the wrong course, they may leave instead of investigating whether another course would be more suitable. I have myself counselled dozens of students who had decided to drop out, but then discovered where their real interests lay, switched to a different engineering subdiscipline, and graduated with high marks. Some of them had been failing miserably until they switched, in spite of having been stellar students at earlier stages of their education.

Finding a mentor can be helpful, although this does not work for everyone, and it is difficult to sustain communication between mentor and mentee over a long period. Creating networking opportunities for students in the first and second years of science or engineering studies to meet students in more advanced years can be successful. There are organisations for women in science and/or engineering in many countries, many of them linked through the International Network of Women Engineers and Scientists (INWES), and they often create opportunities for students to meet women who work in these fields. Some of these groups organize events, conferences, networking and mentoring opportunities, and professional development workshops.

The first-year experience is critical. Some students drop out because they do not receive adequate encouragement and sometimes their career expectations are severely reduced. The academic performance of most students entering the first year of university is lower than they expect from their record at earlier stages, and this can discourage many women. Of course, when students leave, this is also a loss for the university. Some universities have increased levels of retention and morale by providing remedial education for those students, whether women or men, who need help in certain courses.

Increasing the number of women professors can help to provide mentors and role models in the classroom. Some women professors emulate traditional-minded men, but their presence can help nonetheless to demonstrate that women can succeed as well as men. That is one type of message. Although this is not a general rule, potential female graduate students may seek to do their thesis under the supervision of a female professor. There was a time when I supervised twenty students in my laboratory, of whom sixteen were women. I also see female colleagues who have several female graduate students. Some of my male colleagues take on female graduate students as well. There are some research groups that are more women-friendly than others, and potential graduate students should ask students ahead of them about their experiences before choosing a professor or a topic.

Some women faculty work on research topics that have direct benefit for humanity or look at things in a different way.

My own research is on how to predict premature births using only data collected during visits to the obstetrician. The results matched the performance of an invasive testing technique. Another project attempts to assess the level of illness or the effectiveness of therapy in patients with rheumatoid arthritis, using an infrared camera to investigate the temperature of the joints. A third project is to develop strategies to improve the effectiveness of managing medical technologies in developing countries. Over the past twelve years, my research has involved a majority of women students working for masters' degrees, and equal numbers of female and male doctoral students, all of them enjoying the topics and writing interesting theses. Several of the students who finished their masters' degree stayed in my laboratory for their doctoral degree. This provided continuity and mentoring from senior students to newer ones. Our group meetings have also had babies in the room from time to time: those of my students who are mothers are encouraged to bring their babies with them if they wish to during their graduate studies.

In Canada, the United States, and Europe, universities have made efforts to re-examine and improve their policies regarding diversity. The report of the Expert Working Group on Women and Science published by the European Technology Assessment Network (2000, cited in Chapter 8) presented a very detailed overview of the career patterns of female and male researchers in universities, and described several examples of efforts that were being made to increase the numbers of women entering these fields in several member states of the European Union. But policy statements are not enough: the policies need to be explained and enforced for real change to occur. Some student leaders, professors, and administrators have been effective in bringing about improvements in the climate of engineering and science faculties, but one major concern remains: if more women enter these faculties, will they just emulate the men who are already there? Some of the women who have been attracted to these disciplines may be relatively at ease in a predominantly masculine culture, and in any case that culture may influence women's own perceptions and views during their years of study. Later, after women enter the workforce, they may find that they need to establish their credibility and excellence judged by criteria set by the male majority, until there are enough women to create a more inclusive culture.

—⦿—

CHAPTER 12

Women in Scientific
and Engineering Workplaces

Many scientists and engineers would like to assume that they are unbiased, but an analysis of how merit is judged in the world of science and engineering points to a different result. Several studies have demonstrated that, although the selection of faculty, the hiring and promotion of scientists and engineers in workplaces, and the distribution of grants and awards are all supposed to be based on objective criteria, applied to everyone in the same manner, bias and prejudice intervene at every stage. (Of course bias and prejudice affect careers in other fields too, but the focus here is on science and engineering, domains where women are still severely underrepresented.)

In recent discussions of how bias and prejudice can affect career prospects, "diversity" has become a widely used term, although it often goes undefined. Simply put, it refers to those characteristics that make people different from each other, including gender, race, age, ethnic and cultural background, sexual orientation, ability or disability, religion, education, class, marital status or family status, and indeed any other characteristic that shapes an individual's attitudes, behaviour, and perspectives. Respect for diversity means, then, respect for differences among individuals and awareness of how these differences can be involved in interactions between identifiable groups. It also means that when different rules, criteria, or standards are applied to different groups or individuals, in the form of double standards, this can create discrimination.

In dictionaries, "discrimination" is defined, simply and neutrally, as the action of discerning or distinguishing things or people from other things or other people. It is important to note that discrimination can sometimes be positive. For example, providing maternity leave for women expecting babies recognizes the physiological difference between pregnant women and others, and responds to their needs. Paternity leave can also be provided, to enable fathers to share parenting responsibilities once the baby has been born. Another example of positive discrimination is the establishment of means to "level the playing field," as with outreach activities that expose girls and young women to potential careers in fields where women are underrepresented. Hiring women as role models for such activities is also an example of a time-limited form of positive discrimination implemented in order to correct a serious imbalance. Yet another example might be the application of special measures to increase the numbers of men preparing for and entering careers in nursing and other fields of health care.

More often, however, discrimination is negative, concerned with excluding members of different groups from a profession or an institution rather than including them. It has come to be associated with the much more harmful practice of segregation. As we have seen in previous chapters, stereotypes often shape perceptions and can have a major impact on the career progress and success of women and members of other underrepresented groups. We can assume that, in some cases, this happens unconsciously, especially where members of selection and award committees have not been trained to recognize subtle biases or prejudices. However, when negative discrimination, in a place of employment or elsewhere, results from pervasive, interrelated actions, policies, or procedures, it becomes appropriate to regard it as systemic discrimination rather than merely a question of certain individuals' unexamined prejudices. The examples discussed in this chapter reflect the ways in which bias and double standards can be applied in a variety of situations and environments.

THE MYTH OF MERITOCRACY IN ACADEMIA

Margaret Rossiter (1982) describes how women entered science in large numbers in the 1920s and 1930s, did important work as individuals, and even worked on major projects, yet rarely won awards or held office in professional associations. They also had low visibility: women were underrepresented in the higher ranks of academia and the professions and had no prestige. Prize and selection committees passed over outstanding women, either ignoring their work or attributing it to others.

For example, even after the Rockefeller Foundation created and funded postdoctoral fellowships that supported thousands of scientists and scholars, in the United States and in other countries, few went to women (see Rossiter 1982, pp. 269–71). Between 1920 and 1938 in the United States, where the fellowships were allocated by the National Research Council, 395 women obtained PhDs in zoology, but only fourteen of them, or 3.6 percent, won postdoctoral fellowships. Eighty-six women earned PhDs in physics, but only two (or 2.3 percent) received postdoctoral fellowships. In chemistry the proportion receiving such fellowships was even lower, at 0.82 percent, with just four being awarded even though 487 women received PhDs in the subject. Rossiter notes that an unpublished report from the National Research Council recorded these lopsided appointments, but did not even discuss the possibility that an all-male selection committee could have been biased, concluding instead that the female candidates had been weak and undeserving. Women applicants faced a negative reception from officials, who were twice as sceptical of their abilities as they were of male applicants. The women thus had to compete not only against other applicants but also against the selection committee's stereotypes of women's abilities. The proportion of women receiving postdoctoral fellowships in child development was much higher, but, tellingly, these were funded by a separate body, the Laura Spellman Rockefeller Memorial Foundation, and candidates were interviewed by separate selection committees (Rossiter 1982, p. 203).

When recognition is constantly denied, a breakdown in morale occurs. Only a few women broke through the barriers placed against them in the 1920s and 1930s, and usually because they had the enthusiastic backing of powerful male colleagues or, as with the fellowships in child development, were pursuing research in fields sidelined as "feminine." There is evidence that these same types of bias, prejudice, and negative discrimination are still present today, even if they may be less pronounced or pervasive than they were then. For many awards and positions, women need far more evidence than men to obtain the same degree of recognition. This affects many aspects of their careers, from access to academic posts, professorships, awards, prizes, and grants to their prospects for having their work published in leading journals in their disciplines. Sometimes women are ignored, or their work is obscured or appropriated by others. The reality of women's work being less valued than men's is not restricted to women in science or engineering but is a general social reality.

Paula J. Caplan (1993) has discussed the myths in academia that sustain negative discrimination, and in particular the myth of meritocracy. She argues that faculty women in male-dominated fields are more likely than other faculty women to believe in the myth that people are rewarded according to merit—whether by promotion, the granting of tenure, or the receipt of awards or prizes—but, as she writes (pp. 48–49),

> in academia, people are formally rewarded (with a degree, a job, a promotion, tenure, merit increases, or increases in power and status) simply according to the quantity and quality of the work that they do. . . . Many of us find it painful to recognize that making a place for ourselves in the academy can involve putting on shows of various kinds, being in the right places, saying the right things, knowing the right people. We long to have our work speak for itself. . . . The insidious and misguided implication is that all men working in academia were hired solely on the basis of merit, without networks, friendships and so on. A related (and erroneous) implication is that when women are not hired, it is only because their work isn't good enough.

Having more women on selection committees may help, but this is not a certainty, as some women will go against other women in order to show men that they are not feminists or otherwise biased in favour of their own gender. A more permanent solution would be to provide gender-sensitivity training to all members of selection committees, both women and men, and to ensure that there is a critical mass of women on them. Then it may be possible to remove systemic bias from decision-making processes, or at least to minimize its impact.

It should be added that this is a complex issue with several other factors coming into play, such as the wide variation in the level of regard in which different fields of work are held. Women are frequently found working in multidisciplinary fields or in experimental science, areas that may be valued by peers as being less important or prestigious than more abstract disciplines, such as theoretical physics or theoretical mathematics.

A large study by G. Sonnert and G. Holton (1996), assessing the career outcomes of women and men who had received prestigious postdoctoral fellowships from the National Research Council in the United States, brings us back to the main topic of this book. These authors found that, although career outcomes were not very different for men and women in the biological sciences, there were significant differences in outcomes in engineering, mathematics, and the physical sciences. In spite of twenty years of affirmative action in these fields, the "glass ceiling" was still very present. For example, for the 29 percent of women and 27 percent of men who were hired in the top 15 percent of academic institutions in the United States, the average offer to the women was one full rank below the average offer to men (assistant professorships for the women, associate professorships for the men). Nearly three quarters of the women (73 percent) said that they had experienced some form of discrimination through denial of a post or tenure, while 13 percent of the men said that they had experienced reverse discrimination. Women mentioned experiencing subtle forms of exclusion and marginalization. As for opportunities for collaborative work, women said that these often arose when they were to be in subordinate roles, but much less when equal partnership was sought.

A number of the men and the women in this study agreed that gender played a role in their research. Half the women and one quarter of the men said that they thought gender was a factor in how they conducted their research and interacted with other scientists. As to whether gender influenced their choice of research topics, 40 percent of the women and 16 percent of the men agreed that it did, and 36 percent of the women and 20 percent of the men also thought that gender influenced their ways of thinking and their research methods. The women also felt that they were more thorough in preparing their articles and that this resulted in the production of lower numbers of papers. The average number of papers a year was 2.3 for the women, and 2.8 for the men, but the average numbers of citations were 24.4 per paper in women's papers and 14.4 in men's papers. This suggests that measuring the performance and output of academics should involve assessing the quality of work accomplished rather than just the quantity.

Another study, in Sweden (see Wenneras and Wold 1997), assessed a peer-review process that awarded postdoctoral fellowships to candidates with PhDs in the biomedical sciences. The pool of women candidates was 44 percent of the total pool, but women won only 25 percent of the postdoctoral positions and 7 percent of the professorial positions. It was thought for many years that when the numbers of women in a particular discipline increased, male domination of the upper echelons would diminish, but these authors demonstrate that this is not in fact occurring, despite the large numbers of women completing PhD degrees. The authors studied the productivity of the applicants in six different ways: the number of original scientific publications and those in which they were first author; the sum of the "impact factors" of the journals in which the articles were published for each applicant; the total impact factor for articles where the applicant was first author; the number of times the authors' articles were cited in the year 1994; and citations of papers where the applicant was the first author. They found that men and women with equal productivity did not receive the same competence rating from the reviewers at the Swedish Medical Research Council, who gave female applicants lower scores than male applicants. The most productive group of female applicants, with more than one hundred points each, were the only women judged to be as competent as the men, although they were considered to be equal only to the group of men with the least productivity (fewer than 20 points). Independent of their scientific productivity, male applicants received an extra 0.21 points on competence. For females to receive this same additional competence score, they had to exceed male productivity by sixty-four points, which represented three more papers in journals such as *Nature* or *Science*, or twenty extra papers in a journal with an impact factor of three or more. The authors concluded that female applicants had to be 2.5 times more productive than male applicants to receive the same competence score from the reviewers and be awarded a fellowship. Men had been overevaluated and women underevaluated in a systemic fashion.

A second finding of this Swedish study was related to nepotism. In order to be perceived as equally competent, an applicant lacking personal ties with the reviewers needed to have sixty-seven more impact points than an applicant of the same sex who had some association with one of the reviewers. Female applicants lacking personal connections with committee members, and not receiving the male "gender bonus," needed 131 more impact points to be considered as equally competent to male applicants affiliated with one or other of the reviewers. The number of women reviewers, at five out of fifty-five, was too small to assess the impact of having women serving on review committees. This study demonstrates clearly and quantitatively that sexism and nepotism still exist in committees evaluating men and women in science, engineering, and technology.

I participated in several scholarship committees over a period of fifteen years and can confirm the existence of bias. Being frequently the only woman on a science or engineering scholarship selection committee, I observed the following. On a panel of two men and one woman, men chose mostly men as winners, while I, the single woman on the committee, chose a number of excellent women and a number of excellent men prior to the discussion on selecting the final list of winners. I asked what my colleagues' criteria were and then explained mine. After a positive discussion, my colleagues became more open minded, admitted having not seen the qualified women in the same light as I did, and agreed to put some of the women back onto the list of winners. One factor that needs explanation is that in the pool of women who choose non-traditional fields, their average performance is usually higher than the average for the men. For example, the proportion of women who are on deans' lists for excellence is frequently higher than the proportion of men on that list when compared to the pools of women and men students. In each committee in which I participated, the final list seemed to be satisfactory to everyone, and in the following year the same colleagues became better at finding deserving women for the list of winners.

The worrisome point is that there are many selection committees that have never included a woman and have never received gender sensitivity training. Moreover, not all women are aware of gender bias, so having a woman on a committee does not necessarily mean that she will understand potential biases. Women need gender sensitivity training as much as men if justice and a bias-free environment are to be attained. Another point is that the young women who win these scholarships often think that bias no longer exists. They do not know about the work done by committee members that led to their winning the awards they genuinely deserved.

AWARDS, PRIZES, AND MEMBERSHIP OF PROFESSIONAL BODIES

Women are often overlooked or sidelined when it comes to awards, prizes, and membership of scholarly or professional bodies, and even when a woman gets a prestigious award or fellowship, it often comes many years after the work that it recognizes has been done. For example, since the five original Nobel Prizes began to be awarded, in 1901, there have been (as of 2008) 754 male Nobel laureates, but only 35 female laureates, of whom twelve won or shared the Peace Prize, and eleven the Literature Prize, leaving only twelve female winners of scientific Nobels, one of whom (Marie Curie) received two prizes (see Nobel Foundation 2008). (As for the Sveriges Riksbank Prize in Economic Sciences in Memory of Alfred Nobel, introduced in 1969 and sometimes called the Nobel Prize in Economics, no woman has ever won it yet.)

Similarly, although rules formally banning women from membership of some academies and other learned societies were abolished many years ago, the unwritten rule long remained that every member was to ensure that no woman's name was put forward for membership. However, in 1910 Marie Curie was nominated for election to the Académie des Sciences in Paris after winning her two Nobels, first in physics in 1903, for her research on radioactivity, then in chemistry in 1904, for her discovery of radium. Her nomination provoked a heated discussion and a petition against it, organized by Jacques Bétolaud, was brought to the plenary session just before the vote. Then, as Londa Schiebinger writes,

> when Curie's name was finally put to a vote within the Académie des Sciences, she lost to Edouard Branly, a pioneer in wireless communication, by a narrow margin of two votes. But Curie's case raised the more general question whether women should be admitted to any of the great academies of France. This issue was settled by a greater margin: by a vote of 90 to 52, members of the Institut de France [the body that groups five of the French academies] decided that no woman should ever be elected to its membership. Jacques Bétolaud . . . and his *confrères* had carried the day. (Schiebinger 1989, 10–11)

In Britain, where Caroline Herschel (see Chapter 6) had become the first woman honorary member of the Royal Society as early as 1835, it was not until 1945 that two women, Kathleen Lonsdale and Marjory Stephenson, were elected as full members. It was much later, in 1979, that the Académie des Sciences admitted its first woman member, Yvonne Choquet-Bruhat.

Unlike the Royal Society, the Académie des Sciences, and other scientific academies, the engineering academies were, for the most part, created in the latter part of the 20th century, except for the Ingenj rsvetenskapsakademien (IVA) in Sweden, which was established in 1919, and the Akademiet for de Tekniske Videnskaber (ATV) in Denmark, created in 1937. The very first women members were elected to the academies in Denmark and the United States in 1965 (just one year after the latter was founded), and then one woman was elected in Sweden in 1970. According to research by Moyra McDill (2007), the proportion of women members in these academies remains at 15 percent or below:

> At the low end are Japan, [South] Korea, the United Kingdom, and Mexico, at less than 2 percent. In the mid-range of 4 to 5 percent are Canada, Norway and the United States. At the high end are Croatia, Finland, and France at about 10 percent. Sweden stands alone at 15 percent.

As with female Nobel Laureates, women fellows of engineering academies are on average younger than their male counterparts and tend to have made their mark in different areas (see McDill 2007). While 18 percent of male fellows were in electronics, photonics or physics, only 5 percent of female fellows were, but women make up 18 percent of the fellows in interdisciplinary studies, innovation, education, ecology, and history, while the proportion of men in these subdisciplines was only 8 percent.

At the Institute of Electrical and Electronics Engineers (IEEE), the largest technical society in the world, the proportion of women nominated and elected varies greatly from year to year (see Appendix 3). Between 1999 and 2009, the highest proportion of women elected came in 2008, when there were 27 women out of a total of 295 new fellows (or 9.2 percent of the total): 47 women had been nominated, making the success rate 57.4 percent. The next most successful year was 2007, when 8 out of a total of 268 new fellows (or 6.7 percent) were women: 48 women had been nominated that year, so the success rate was only 37.5 percent. The year with the highest success rate was 1999, when 61.9 percent of the women nominated were elected (thirteen fellows from twenty-one nominations), but the proportion of women elected in 1999 was only 5.3 percent. The lowest number of nominations was in 2000 (six women nominated and two elected), so that 0.8 percent of new fellows that year were women. Evidently, a larger number of nominations of women does not necessarily mean that more are being elected. For example, in 2004 thirty-six women were nominated and only six were chosen. However, a good approach remains to nominate as many qualified women as possible in any given year.

DISCRIMINATION IN PAY AND JOB EVALUATIONS

Around the world, and in almost every workplace, women still tend to be paid less than men for similar work. Women workers also tend to be concentrated in fields regarded as "pink collar," which are less highly valued than the white-collar and blue-collar jobs dominated by men. Arguments against equal pay for women often use job differentiation as the basis for justifying higher pay for men. What this argument ignores is the value that society puts on these jobs, which is not necessarily what their true value should be.

Canadian statistics on average levels of pay by gender are similar to those found in the United States and Europe. Comparing the pay of secretaries or administrative assistants with the pay of gardeners or technicians, we can observe how each type of work is valued. It is clear that jobs regarded as "masculine" are better-paid and valued more highly than jobs that are regarded as "feminine."

One of the early studies that compared evaluation of women's and men's work was done by Philip Goldberg (1968). He asked both women and men to evaluate essays for quality and strength but switched the names on the essays so that those written by men had female names cited as authors, and vice versa for those written by women. He found that both women and men rated the essays attributed to men but written by women as superior to the essays attributed to women but written by men. The evaluators, of both genders, sincerely thought that the essays presented to them under male names were more important, more authoritative, and more convincing. This study was repeated by M. A. Paludi and L. A. Strayer (1985), who asked 300 college students, 150 of each gender, to evaluate academic articles in the fields of politics, psychology of women, and education. These topics were assumed to be classified as masculine, feminine, and neutral, respectively, and presented as having been written by authors with male, female, or neutral-sounding names. Again, the articles apparently written by male authors were judged more positively. In the case of neutral names, the subjects' bias against women was stronger when it was assumed that these authors were female.

A study of the success, failure, and ability of university administrators (originally reported in an unpublished monograph in 1986, and then in Haslett, Geis, and Carter 1992) found that if a male administrator frequently failed to complete his work, the perception was that the job was too heavy for him and that he needed an assistant, and thus more responsibility and salary. However, if a female administrator frequently failed to complete her work, the perception was that she could not handle the job and needed to be placed in a less responsible and lower-paid position.

In a large study of scientists and engineers in high-tech companies in the United States, conducted by Nancy DiTomaso and George Farris (1992), Caucasian men were rated by their managers as average on the attributes of innovation, usefulness, and promotability, and rated a little lower on cooperativeness. These managers rated women lower on all these attributes except cooperativeness. When the employees recorded their own self-assessments, however,

Caucasian men rated themselves slightly higher than their managers did, while women rated themselves lower for all attributes except cooperativeness. One possible interpretation of this result is that Caucasian male scientists and engineers understand the corporate culture better and interpret the feedback more accurately, since they come from backgrounds similar to those of their managers. For women, self-confidence may be an issue and the uneven understanding of feedback may be a problem. The lower self-esteem of teenage girls may continue into later life for some women, especially when they are working in non-traditional occupations, where the standards of achievements are likely all in male terms.

In the same study, DiTomaso and Farris report that the women rated their managers lower than men did on the following skills: getting people to work together, letting people know where they stand, being sensitive to differences among people, and minimizing hassles with support staff. However, the women rated managers more highly on communicating goals clearly, defining problems, getting resources, and motivating commitment. This study suggests that managers need to put more effort into developing objective and measurable criteria for the assessment of all their employees, and focus more attention on the type of feedback they provide. They should also be more sensitive to diversity issues and make sure that the assessment criteria and rules are well understood by everyone. Managers should also make efforts to better understand the different approaches and perspectives that women (and members of minority groups) can bring to the organization, and avoid underestimating the performance of particular groups of employees based on stereotypical visions of their potential. If managers build teams with people from diverse backgrounds and perspectives, their organization's performance can be expected to improve.

Research by Martha Foschi, Larissa Lai, and Kirsten Sigerson (1994) confirms that discrimination applies in relation to temporary positions as well as permanent ones.

Their study of the hiring of students for summer positions showed that even though groups of female and male candidates' files had been made equal as to their background and skills, the men were viewed as better qualified by a majority of the male subjects in the study, who chose sixteen out of twenty men as successful candidates. Only four women were selected as most qualified. This bias was even more pronounced when the field of the candidate was one considered non-traditional for women. Foschi and her co-authors also showed that when women participants made the hiring decisions, they rejected the stereotypes by selecting close to half of the women candidates as winners: eleven out of twenty female candidates were chosen by female selectors, and women selected nine men out of twenty as most qualified. This study shows that men still demonstrated gender bias in the choices they made, while women seem to have been more capable of fairness.

DIFFERENCES IN MEN'S AND WOMEN'S CAREER PATTERNS

The story of the two young male professors who believed that women who entered graduate school should not have children (see Chapter 11) indicates that there is still a belief in the myths that women become less committed to their studies or careers after they have become mothers, and that it is not possible to balance a family and a career. However, starting a family while enrolled in graduate school is likely less stressful than doing so when starting a new job, and no matter when child-bearing occurs, our institutions should support this choice. It makes sense both economically and socially. Women, and men, who have spent years in demanding studies will not abandon their careers to start a family. Moreover, it is far less disruptive for an organization for one of its employees to have a baby than if an employee has a serious medical emergency. Pregnancy is not a disease, and pregnant women are generally healthy and can stay connected with their workplaces. Medical problems such as heart attacks, mononucleosis, depression, and alcoholism can be far more disruptive when planning work assignments and responsibilities than planned maternity or paternity leave.

Felice Schwartz and Jean Zimmerman (1992) have presented the results of an interesting study of the patterns of work for men and women in the United States in 1962 and in 1992. They show that in 1962, in an era when the fertility rate averaged 3.7 births per woman, the average woman worked full time between the ages of twenty-two and twenty-five, took a career break to raise children between twenty-five and thirty-five, worked part-time between thirty-five and forty-five, and only then (if at all) returned to full time work, which lasted until her retirement. In contrast, the average man worked full time, without career breaks, from the time he left the education system until he retired. The picture for both women and men looked quite different thirty years on. By then, the average woman moved into part-time work between the ages of thirty and thirty-five, but had full-time jobs for the rest of her working life, while the average man also worked part time for five years, but later, between sixty and sixty-five. Against this background, Schwartz and Zimmerman (1992, pp. 55–70) explain why companies should make efforts to hire and retain professional women:

> Before she has a baby, today's woman has chosen a career, trained for it, gained substantial experience, and given her employer ample time to assess the quality of her performance. By that time, if she's good and seriously motivated, she is a highly valued, seasoned professional in whom the company has made a substantial investment.

Rather than doing this, however, many managers appear to project their own experiences, and their traditional lifestyles and beliefs, when making decisions in the workplace, and fail to take account of the changes occurring today in gender roles, such as the increased participation of men in caring for children and sharing household duties with their working partners, and women's serious commitments to their careers. Companies failing to recognize the need for flexible policies that allow young parents to balance families and careers will eventually lose their best employees, male and female, who will seek posts at more progressive firms. False assumptions about gender roles will only disappear when companies hire capable women who, through commitment and experience, can succeed in dispelling such damaging views. When men too demand more flexibility, as they take their fair share of parenting responsibilities or seek to balance lifestyle with work, the process of change will accelerate.

In 1998, the Ordre des Ingénieurs du Québec, an association for professional engineers in that province, conducted a survey that included questions on the amounts of time taken off work by male and female engineers. The survey found that the average length of leave was not significantly different for women and for men, but their reasons for taking leave differed. Men had taken an average of three weeks for parental leave and the women several months. Other men took long leaves for reasons related to their lifestyles. Interestingly, a previous study by a large engineering employer in 1991 (see Canadian Committee on Women in Engineering 1992) showed that parental leaves were taken only by women. Thus, in the space of seven years, several more men were seen to participate in parental care. The same firm also ruled out maternity as a factor in women leaving their jobs.

Some women engineers do drop out, especially if they have been working in hostile environments, and this may happen to coincide with pregnancy in some cases, but the main factor in their leaving is usually the poisonous or negative atmosphere at the workplace. Several such women have told me that they felt they might as well take care of their babies until they could find better employers. After taking career breaks, these women often re-entered the job market at different firms.

THE IMPACT OF SEXUAL HARASSMENT IN WORKPLACES

Having discussed sexual harassment in schools (in Chapter 9), we now turn to the question of its impact in workplaces. Bernice Sandler and Robert Shoop (1997, pp. 4–5) define sexual harassment as follows:

> Unwelcome sexual advances, requests for sexual favours, and other verbal or physical conduct of a sexual nature constitute sexual harassment when any one of the following is true: submission to such conduct is made either explicitly or implicitly a term or condition of a person's employment or academic advancement; submission to, or rejection of, such conduct by an individual is used as a basis for employment decisions or academic decisions affecting the person. Such conduct has the purpose or effect of unreasonably interfering with a person's work or academic performance, or creating an intimidating, hostile, or offensive working, learning, or social environment. . . . The behaviour is unwanted or unwelcome; the behaviour is sexual or related to the sex or gender of the person; and the behaviour occurs in the context of a relationship where one person has more formal power than the other, or more informal power.

Formal power relationships include, for example, those between a work supervisor and an employee, a faculty member and a student, or a physician and a patient, while an informal power relationship exists where one peer or colleague, while nominally equal, exerts an influence over another.

The frequency of sexual harassment in engineering workplaces is suggested in a report, specifically on chemical engineering, by Claudia Caruana and Cynthia Mascone (1992), who found that at least 67 percent of women in factories and 57 percent of women in offices had suffered some type of harassment during their careers. Those women who had been harassed had different perceptions about harassment than those who said that they had never been harassed. For example, 51 percent of women who responded that they had been harassed thought that it was a common occurrence, 47 percent felt that it was occasional, and 2 percent thought that it was rare. On the other hand, only 4 percent of women who said that they had never been harassed thought that it was common,; 69 percent thought that it was occasional, and 27 percent saw it as rare. These results indicate that many women, at least at the time Caruana and Mascone did their research, had little knowledge about sexual harassment. I have myself met several women who were harassed before the 1990s but did not understand that what happened to them was harassment. For example, one colleague said that she had been forcefully kissed by her boss one day, but she just thought that that was what some men do, so you just had to put up with it and hope that it was not repeated.

Another obstacle to understanding the nature of harassment is that it frequently occurs for reasons other than seeking sexual favours, which is how the term is widely understood, on a narrow definition that tends to be reinforced by media coverage (see Frize 1995). Harassment is a question of power and domination. It can take the form of a backlash, intended to disorient women or discourage them from working in a male-dominated environment. Treating women as sexual objects or making sexist remarks in their presence may have nothing whatever to do with sexuality as such, but instead is deployed to make women feel that they do not belong in a particular workplace or industry.

I have counselled several women graduate students who had been sexually harassed by their supervisors, but felt helpless because of their vulnerability regarding their thesis work, and blamed themselves for the situation. One of these students had the courage—and it takes tremendous strength to go through with this process—to make a formal complaint against her supervisor. She managed to find a new supervisor and eventually, after some delay, defended her thesis successfully. The fact that a complaint can take several years to process is emotionally taxing for both victim and predator, but she wants this to act on behalf of students who come after her.

Sexual harassment is clearly not a problem that faces only women engineers or scientists, but it may be more difficult for them to cope with. Because of the small number of women at the site or in the firm, they may find less support and fewer response mechanisms than if they were working in firms where there are many professional women. Moreover, women may find little support or sympathy in a climate of "Boys will be boys," mainly in firms that are male dominated. Policies and procedures for dealing with harassment also vary from one company to another. Education is at the heart of the solution (see Frize 1995). To eliminate harassment in workplaces, companies must maintain fair investigation procedures that do not victimize complainants and provide checks and balances for adequately verifying accusations. Providing moral support to colleagues who are complainants can help to reduce the stress they feel. Informal investigations can often accomplish more than formal ones, which are more confrontational. However, in some situations, a formal complaint is the only possible approach. Organizations should provide both formal and informal mechanisms when they are developing policies and procedures.

IMPROVING THE STATUS OF WOMEN IN WORKPLACES

A report prepared for the Massachusetts Institute of Technology (MIT) (1999), building on earlier studies to investigate the status of women faculty in science at that prestigious institution, contains some thought-provoking observations:

> A common finding for most senior women faculty was that the women were "invisible," excluded from a voice in their departments and from positions of any real power. This "marginalization" had occurred as the women progressed through their careers at MIT, making their jobs increasingly difficult and less satisfying. In contrast, junior women faculty felt included and supported in their departments. Their most common concern was the extraordinary difficulty of combining family and work. An important finding to emerge from the interviews was that the difference in the perception of junior and senior women faculty about the impact of gender on their careers is a difference that repeats itself over generations. Each generation of young women, including those who are currently senior faculty, began by believing that gender discrimination was "solved" in the previous generation and would not touch them. Gradually however, their eyes were opened to the realization that the playing field is not level after all, and that they had paid a high price both personally and professionally as a result.

The report offers a number of suggestions to eliminate gender bias, compiled from research and discussions undertaken between 1995 and 1999, notably including the following, from a list of recommendations made by the Committee of Women Faculty in the School of Science in 1996:

> Maintain and open channels of communication between department heads and women faculty.
>
> Collect equity data each year for inclusion in a written report, and disseminate a summary of the report to the MIT community.

Raise community consciousness about the need for equity.

Seek out women for influential positions within department and Institute administrations, including as heads and as members and chairs of key committees. Involve tenured women faculty in the selection of administrators, and consult with women faculty to ensure the continued commitment of administrators to women faculty issues.

Review the compensation system, which has been shown to impact differentially on salaries of men and women faculty in recent years. In particular, review the reliance on outside offers. Review salary data and distribution of resources annually for gender equity.

Replace administrators who knowingly practise or permit discriminatory practices against women faculty. Promptly end inequitable treatment of women faculty, and make appropriate corrections for inequities when they are discovered.

Watch for, and intervene to prevent, the isolation and gradual marginalization of women faculty that frequently occurs, particularly after tenure.

Take proactive steps . . . to promote integration, and to prevent isolation of junior women faculty.

. . . Make the policy on maternity leave and tenure clock uniform throughout the Institute, and make the policies widely known so that they become routine.

Take steps to change the presumption that women who have children cannot achieve equally with men or with women who do not have children.

Advise department heads to place senior women faculty on appropriate search committees.

. . . Inform department heads each year that conscious effort is needed to identify and recruit outstanding junior and senior women faculty from outside MIT.

Address the family–work conflict realistically and openly, relying on advice from appropriate women faculty, in order to make MIT more attractive to a larger pool of junior women faculty, and to encourage more women students and postdoc[toral fellow]s to continue in academic science.

These findings and recommendations have a relevance that goes far beyond MIT alone. Carefully adapted, they can be of value for any university, or any other workplace, that wishes to improve the status and the numbers of women qualified in science, mathematics, technology, or engineering.

—∞∞∞—

CHAPTER 13

Strategies for Equitable Workplaces

The "gender gap" in employment is the subject of a recent study by an economist at Goldman Sachs, Kevin Daly (2007). Daly points out that "in labour force survey data across countries, the child care requirement is consistently the most commonly cited reason for female inactivity in the labour force" (p. 9) and argues that if levels of employment for women equalled those for men, gross domestic product could be raised by 21 percent in Italy, 19 percent in Spain, 16 percent in Japan, and 9 percent in Germany, France, and the United States (p. 5). Closing the gap between levels of male and female employment would also, he argues (p. 3),

> help to boost low fertility rates. . . . In countries where it is relatively easy to work and have children, female employment and fertility *both* tend to be higher. It is no coincidence that Italy and Japan have both the lowest levels of female employment and the worst demographic prospects.

Daly's findings support the notion that women can have both careers and children, and that their doing so can be good for the economy. Strategies to support the retention of women, and men, in the workforce while they are raising their families are key to the successful integration of women into workplaces in general, and scientific and engineering workplaces in particular.

HIRING MORE WOMEN IN SCIENCE AND ENGINEERING FACULTIES

When I was the only woman student in an engineering faculty in the 1960s, there were times when, in the middle of an electrical engineering class, I would ask myself: "What am I doing here?" However, I was determined to show that women could do the work. The pressure to succeed was enormous, because if I failed, it would be easy for the professors and the students to think that women did not belong among them. I found the material dry and somewhat boring at times, but because I enjoyed solving problems and wanted to help society, I persevered. Unfortunately, my first job, at a telecommunications company, did not bring me much satisfaction, so I stayed for only one year and then went overseas to study biomedical engineering. There has never been a boring day in my career since then. The conclusion I draw from my own experience is that engineering can be fascinating, but everyone has to find their own niche, the area where his or her interest lies. There are many choices of career paths, and not always in technical streams. Some engineers become patent agents, others are involved in technical writing, yet others can get involved in marketing and sales, safety, standards, design, client support, or one of many other areas.

Professors can be of great help in counselling students about their choices in undergraduate and graduate studies, and their career paths, and in addressing the question of balancing personal life and work. In particular, they can help students to find where they fit best. For these purposes, having more women teaching and doing research in science and engineering faculties would provide role models and potential mentors for women undergraduates and graduate students. Although women faculty are not always aware of gender issues, they could still be good role models if they demonstrate competence and offer mentoring to students. Some students have never had a woman teaching them in these subjects, reinforcing the idea that such studies are only for men, and they may even question their own presence in these faculties.

JoAnn Moody (2004) presents some of the myths and excuses used to avoid hiring new faculty from underrepresented groups, and recommends good practice for university presidents, provosts, deans, and academic departments. She suggests instituting a continuous, year-round recruitment process, accompanied by coaching and monitoring of search committees to ensure that their members understand equitable practices and comply with them. She also recommends measuring whether outcomes show progress towards increases in hiring from underrepresented groups; appointing a diversity committee; holding deans and department chairs accountable for increasing diversity; assisting faculty with spousal job hunting; and paying attention to the lifestyle concerns of candidates (Moody 2004, pp. 90–104). She also offers eight forms of good practice for academic search committees to help them in avoiding sloppy, biased thinking and decision-making, snap judgments and pretexts, and elitist behaviour, such as undervaluing the institutions from which candidates obtained their degrees (Moody 2004, 106–08).

One argument that frequently comes up in search committees in male-dominated departments is "But we cannot lower our standards." This usually suggests that hiring a woman or a person from a visible minority will necessarily do that. Yet it is frequently observed that the bar is actually raised for these candidates, when compared to the expectations of candidates who come from the majority group, usually white males. It is pertinent to review the criteria for judging achievement and success that affect decisions in hiring, tenure, promotion and the awarding of research chairs. Universities must create policies that allow young faculty members, both women and men, to balance family and career. The criteria used to assess faculty performance, based on decades of tradition, need to be re-examined to see if they are still relevant in a changing world. This applies not only to how merit is defined and measured but also to how awards, appointments, and promotions are allocated. For example, examining the quality of publications, rather than their number, and looking at the future potential of candidates as well as what they have accomplished by the time of their interviews, would be fairer to those candidates who have taken career breaks in order to have children. Outdated stereotypes and biases can affect success rates in all forms of competition. These biases can be reduced through education and sensitization, by ensuring that there is a fair gender representation on the committees making decisions, and by making proactive efforts to find qualified women for positions or awards.

ATTRACTING WOMEN INTO SCIENTIFIC AND ENGINEERING WORKPLACES

The first step in finding solutions is to recognize the challenges faced by women scientists or engineers and develop effective structures to address them. Whether a workplace is hospitable and equitable for women depends very much on the attitudes and practices of its management team, and on how well its equity policies are applied and enforced. Some firms make it clear enough that they do not want to hire women— even though this is illegal in most western countries—while others make a point of being proactive about hiring and promoting qualified women.

The development of policies guiding the hiring, promotion, and behaviour of employees is an important step, but no policy can be effective unless it is explained clearly to staff and managers, and enforced. Creating a committee that identifies issues affecting minority groups within an organization can provide many long-term solutions. When issues arouse anxiety or anger in a group, such feelings can be diffused and eliminated by an open and frank dialogue between the diverse groups and with management.

Realistic hiring objectives should be set, based on the availability of women in the pool of graduates. Objective hiring criteria must be established, jobs posted, and women proactively sought. Training should be available to sensitize staff involved in the employee selection process, so that they can recognize which questions are appropriate and which are not, and treat every applicant with equity and respect. One way to test the appropriateness of questions put to women applicants in interviews is to ask oneself whether a man would be asked the same question. If the answer is no, it should not be put to anyone. Each organization has to set its own goals, based on where it currently sits on the scale of progress, and draw up an achievable plan to create a workplace that is comfortable for everyone.

Additionally, to attract more women to apply for positions employers need to make special efforts to encourage women to consider these opportunities. It is not sufficient to advertise the jobs in traditional media. To attract and retain qualified women, an employer needs to identify potential candidates and demonstrate to them that the organization has family-friendly policies, fair hiring and promotion practices, and a policy of zero tolerance of sexual harassment; that it provides regular training for managers on avoiding gender stereotypes and biases; and that gender-awareness training is provided to all its employees. Ensuring that there are women who are aware of potential gender discrimination on the hiring and promotion committees also helps.

WHY WOMEN LEAVE

Anne Van Beers (1996) found in her study of twenty female and twenty male engineers in the Vancouver area that the reasons most cited by the women for leaving employment were that the work ethic was too rigid, there was a lack of flexible work options, they had been negatively affected by the "old boys' network" or by harassment, or they had been made to feel that they were not involved in decision-making processes. In contrast, the men said that they left because of the poor status of the profession, low salaries, or lack of job stability. The issue of career progression seems to have the most damaging effect on the retention of women. In order to increase the pool of qualified women at higher levels, women should be fast-tracked: that is, women with potential should be identified, provided with management training, and promoted to an appropriate level. These women also need to be supported by higher management if this action is to be successful.

However, it also happens that women who are offered the opportunity to become managers turn it down because they want to have more balance in their lives. They can observe that people in management positions do not appear to have much time for their personal lives and conclude that it would be too much of a sacrifice, for themselves or for their families. On the other hand, many men in management positions have wives at home supporting their careers by taking care of all the parental and household chores, thus enabling them to spend most of their time at work. This is not an option for most women, as they still bear the brunt of work in the household. Another avenue is for women not to have children, but this is unacceptable to many women who plan to have families, as well as for society as a whole. In the end, it is the corporate world that needs to change to allow both women and men to have satisfying careers in balance with their personal lives. The rising generation now entering the workforce has expectations about their need for more time away from work and employers will have to adapt if they want to hire them. This will become critical as the "baby boomers" retire in large numbers.

Flexibility in working hours during the relatively few years when employees have young children can generate substantial financial benefits in the long run, reducing staff turnover and thus the cost of hiring and training new people. Parental leave should be available to both mothers and fathers. Having flexible time for work and accepting a certain amount of working from home using electronic communications will definitely help to retain women once they begin to have children.

Access to affordable child care is a major factor in retaining young parents in today's workplace. However, employers need to consider more than just the care of children, since many families today also need to provide care for elderly or disabled parents or relatives. It is important to ensure that women and men can maintain their competence and skills while away on extended leave.

Another point is to encourage lifelong continuing education for all. Employers should provide opportunities for a network to develop and coach new staff. If they build teams with diverse backgrounds and perspectives, results will be improved. Another strategy is to give visible assignments to people who need to build their self-confidence and credibility within the organization.

WHY WOMEN STAY

In her study, Van Beers (1996) reported that the factors encouraging women to stay in engineering included their enjoyment of problem-solving and breaking new ground, their awareness of being role models, and their desire to make a difference. To succeed, women in her study said, you have to have a sense of humour, tough skin, stubbornness, a willingness to take risks, and the capacity to be outspoken when necessary. On the other hand, the characteristics that these women cited as drawbacks in the profession included being too honest and approachable and being outspoken about the environment or gender issues.

Retaining women also depends very much on establishing a support network so that women can learn from each other, share their experiences, and understand not only how to survive but how to succeed. Mentoring younger women provides excellent opportunities for older women to share their expertise and work contacts and play a part in the empowerment of more women. As the number of active women in the field increases, many barriers may fall.

Since the beginning of the 1990s, a number of employers have made visible efforts to hire and promote more women and to recognize and value their contributions. Some have come to realize that if they make changes to attract and retain talented women, they will also continue to attract and retain talented men. Today, the job market in certain fields is competitive, and employees can be more demanding about working conditions and the quality of the workplace environment. This will contribute to making these workplaces more human and allow employees to have lives outside their work. Those employers who understand these principles will find and keep the human resources they need to remain competitive.

In the entrepreneurial sector, we need to change the perception that women are merely moving into engineering positions created by men. The creation of technically based firms by women must be supported by financial incentives for those who would be willing to consider leaving their existing jobs to create such businesses.

The benefits of hiring women will only accrue if women are not simply emulating men, and if they feel that feminine values and attributes are valued. Some women may go out of their way not to be seen to be helping women, and some young women do not want help from other women in any case. If positive change is to occur, the commitment of the women who are promoted into management is essential, and young women need to be enlightened about the damage caused by the denial that gender issues exist.

WORKPLACE CULTURE AND WOMEN ENGINEERS IN THE AEROSPACE AND HIGH-TECH INDUSTRIES

The culture in various engineering workplaces differs according to the type of business. In their study of the occupational experiences of a cohort of male and female engineers, G. J. Robinson and J. S. McIlwee (1991, p. 406) observed that success or failure was partly related to how well the engineer manoeuvred through the workplace culture:

> Women are least occupationally mobile where the culture of engineering dominates. Where engineers as a group are powerful, workplace culture takes on a form strongly identified with the male gender role, emphasizing aggressive displays of technical competence as the criteria for success. It thus devalues the gender attributes of women and equates professional competence with "masculinity." Conversely, where engineers enjoy less power, the workplace culture less closely reflects their interests, to women's benefit. Women's odds of success improve as competence is less male-defined.

According to Robinson and McIlwee, the lowest echelons in engineering workplaces are in sales and marketing, as they are outside the technical hierarchy. Below design are the activities that support design, and above design are the senior and project engineers. These authors found 15 percent of the women and 20 percent of the men in their study in positions above design. In design itself, they found a little more than

40 percent of the women and close to 60 percent of the men, even though the women were technically as proficient as the men, and their grade point averages in engineering were equal to, or exceeded, the men's. The women frequently cited mathematical skills as grounds for their career choice (80 percent of the women, as against only 18 percent of the men), and pursued graduate studies in very similar proportions (25.3 percent and 23.3 percent respectively). Robinson and McIlwee concluded (1991, p. 408) that

> engineers show no significant gender differences in work-related values such as desire for occupational autonomy and participation in strategic projects, attitude toward creative technical work, and interest in acquiring administrative experience. . . . Other factors must explain the occupational disparities.

In particular, their study looked at engineers in the aerospace industry (34 percent of their sample) and in high tech (28 percent). They found a significant divergence between men and women in high tech, but none in aerospace, and pointed out that the aerospace firms were large, had government contracts, and were more affected by affirmative action guidelines. They also had a traditional bureaucratic structure, so that job descriptions, paths of authority, and advancement criteria were standardized and clear. In contrast, high-tech companies tended to have informal and ambiguous organizational structures, in which (Robinson and McIlwee 1991, p. 409)

> rapid growth and the importance of innovation to survival make change constant, job assignments vague and open-ended, authority relations nebulous, and formal communication channels rare. . . . Personal reputations and peer evaluations count more for advancement than formal requirements.

It is not difficult to see why women would fare better in large firms where engineers have less power, and employment equity and clear guidelines are in place, than in small firms where the masculine culture reigns, and peer assessment of women's performance and attributes could be expected to be fairly negative.

PROMOTING WOMEN TO
LEADERSHIP ROLES

Catalyst, a non-profit organization based in New York, but with branches in Canada and Europe, works under the slogan "Expanding opportunities for women and business." In 2007, it published a report arguing that gender stereotyping is "one of the key barriers to women's advancement in corporate leadership, and leaves women with limited, conflicting, and often unfavourable options, no matter how they choose to lead" (see Catalyst 2007).

The report notes that, "even though women make up over 50 percent of the management, professional, and related occupations, only 15.6 percent of *Fortune* 500 corporate officers and 14.6 percent of *Fortune* 500 board directors are women." It points out,

> Women leaders are perceived as "never just right." If women business leaders act consistent with gender stereotypes, they are considered too soft. If they go against gender stereotypes, they are considered too tough. . . . Women leaders face higher standards than men leaders and are rewarded with less. Often they must work doubly hard to achieve the same level of recognition as men leaders for the same level of work, and "prove" they can lead. . . . When women exhibit traditionally valued leadership behaviours, such as assertiveness, they tend to be seen as competent, but not personable or well-liked. Yet those who do adopt a more stereotypically

feminine style are liked, but not seen as having valued leadership skills.

Based on survey data from the United States and Europe, the report examines the effects of gender stereotyping in workplaces, the consequences of gender bias, and three specific "double-bind dilemmas" experienced by women business leaders. It also suggests solutions to counter the persistent effects of gender stereotyping. Raising awareness of how stereotypes operate and making individuals accountable can decrease the negative effects of bias. To this end, the report recommends

> providing women leaders and all employees with tools and resources to increase awareness on women leaders' skills and on the effects of stereotypic perceptions; assessing the work environment to identify in what ways women are at risk of stereotypic bias; [and] creating and implementing innovative work practices that target stereotypic bias. These practices can be particularly effective when they address specific areas of risk, such as in an organization's performance management procedures.

In particular, the report suggests that organizations should provide training and diversity education for their managers on bias and its consequences, inconsistencies between values and actual behaviour, the causes and effects of gender inequality in workplaces, and the need for objective and clear performance and evaluation criteria.

Resources to provide this training can be found not only on the Internet, but also through consultants who run workshops on gender equity in workplaces. One such resource, developed and made available on the internet by Virginia Valian, with the support of a grant from the National Science Foundation in the United States, comprises four tutorials in the form of scripts with voice-over narration, explaining gender schemas related to science careers (see Valian 2007).

SUPPORT FROM PROFESSIONAL ASSOCIATIONS AND SCIENTIFIC SOCIETIES

Women continue to represent only a very small proportion of professional engineers, in Canada and elsewhere. According to Engineers Canada/Ingénieurs Canada, the national federation of regulatory and licensing associations for the profession, in 2002 only about 9 percent of the 160,000 engineers in Canada were women. The concentration of women in a small number of subdisciplines is also a feature of the profession in most western countries: women tend to be present in higher proportions in environmental, chemical and biomedical engineering than in mechanical, electrical, computer, and software engineering.

Progress in professional engineering associations or scientific and technical societies can be measured by monitoring the proportion of women on their executive committees, the pattern of awards and recognitions given to men and women, and the numbers of women invited to their conferences as keynote speakers, panellists, or plenary session speakers. When the proportion of women in each of these activities reaches or surpasses their actual proportion as members of an organization, then progress is visible and real. Deserving women can be found if one looks for them, and recognizing their achievements and expertise will make them feel that they form an integral part of the profession, as partners in designing tomorrow's world.

⊸∞⊷

CHAPTER 14

Developing a New Culture in Science, Engineering, and Technology

As we saw in Chapter 4, historically there has been a close association between views of women as being "naturally" incapable of engaging in science or engineering and the use of metaphors of nature, or knowledge, as a woman to be wooed or controlled by male scientists. This association continues to be made even today. In 1965, for example, at the beginning and the end of the address he gave after receiving the Nobel Prize in Physics, Richard P. Feynman (1972) referred to the theory for which he won the prize in the following terms:

> I fell deeply in love with it. And, like falling in love with a woman, it is only possible if you do not know much about her, so you cannot see her faults. The faults will become apparent later, but after the love is strong enough to hold you to her.
>
> . . . So what happened to the old theory that I fell in love with as a youth? Well, I would say it's become an old lady that has very little attractive left in her, and the young today will not have their hearts pound any more when they look

at her. But we can say the best we can for any old woman, that she has been a very good mother and she has given birth to some very good children.

Remarkably, few commented at the time on this distinguished physicist's use of such ancient stereotypes. Eighteen years later, however, there was understandable controversy when Vivian Gornick (1983, pp. 36–37) reported the views of another Nobel Laureate in physics, Isidor Isaac Rabi (who had won the prize in 1944): he told her that women were unsuited to science because of their nervous systems, and declared, "Women may go into science, and they will do well enough, but they will never do great science." It is not difficult to infer that Feynman and Rabi each felt comfortable making these remarks because they knew that many male scientists would agree with them.

When feminists and social critics speak of the dichotomies of gender, they are in part referring to the stereotypes that arise from, and feed into, just such comments, uttered throughout the centuries by men and even by some women. Evelyn Fox Keller (1985) has pointed out how science and "higher," abstract thinking have been generally associated with everything on the list of masculine attributes, and taken to be the exact opposite of what are claimed to be feminine attributes. Linda Jean Shepherd (1993, p. 5) has noted how society still tends to perceive gender attributes today (specifying western culture, though many non-western cultures are marked by similar attitudes):

In western culture, the successful man is considered to be objective, intelligent, logical, active, rational, independent, forceful, risk-taking, courageous, aggressive, competitive, innovative, and emotionally-controlled. . . . Western society expects women to be nurturing, receptive, emotional, irrational, intuitive, subjective, compassionate, sensitive, kind, not aggressive, and uncompetitive.

Shepherd also suggests (1993, p. xiv) that this is one reason why women scientists are reluctant to express feminine qualities in their work, fearing loss of credibility, and that some women scientists are nervous about being labelled as "feminine" because of the potential attachment of the adjective to the concept of biological determinism.

FORMS OF EXCLUSION

The belief that certain specifically "masculine" attributes make good scientists, while "feminine" attributes form an obstacle to scientific achievement, has been transmitted century after century, contributing to the persistence of the male-dominated culture of science and engineering to this day. Linda Jean Shepherd (1993, p. 52) cites the interesting work of Ian I. Mitroff and his colleagues, who presented lists of paired attributes to scientists and asked them which attribute in each pair best described their activities. Most of the respondents affirmed that the ideal scientist would be aggressive, hard driving, self-serving, power oriented, authoritarian, sceptical, diligent, and precise, and 31 percent of them refused to choose between the attributes "warm" and "cold" because they felt that these terms were irrelevant to science.

The male domination of science and engineering may even have been reinforced after the industrial revolution, when these activities, formerly dismissed as crafts or pastimes that women or lower-class men could take part in, increasingly adopted industrialized methods of production, in which knowledge is frequently seen as the result of experimentation using giant machines or involving huge teams of researchers. One example of the former is the particle accelerator, whose cost is so immense that these machines are restricted to the United States, Russia, Japan, and the shared European facility at the Centre Européen pour la Recherche Nucléaire (CERN) in Switzerland. An example of the latter is the Human Genome Project. In these and other cases, experiments now need large numbers of personnel and substantial funding, and results may be published in papers that each have dozens of authors, but may or may not acknowledge the support of the technical staff without whom the experiments would not take place. Class also plays a part. As the British sociologist Michael Young observed in 1971, "What does and does not count as science depends on the social meaning given to science, which will vary not only historically and cross-culturally, but within societies and situationally."

In another context, traditional knowledge, especially that created in Latin America, Africa, or Asia, has been largely ignored by western science. Vandana Shiva (1989, p. xvii) equates the scientific revolution with the rise of a patriarchal science of nature, and the industrial revolution with a patriarchal mode of economic development in industrial capitalism:

> The scientific revolution in Europe transformed nature from *terra mater* [Mother Earth] into a machine and a source of raw material; with this transformation it removed all ethical and cognitive constraints against its violation and exploitation. The industrial revolution converted economics from the prudent management of resources for sustenance and basic needs satisfaction into a process of commodity production for profit maximization. . . . modern science provided the ethical and cognitive licence to make such exploitation possible, acceptable, and desirable. The new relationship of man's domination and mastery over nature was thus also associated with new patterns of domination and mastery over women, and their exclusion from participation *as partners* in both science and development.

A different path might be followed by incorporating the concepts of sustainable development in future work. It is encouraging to see that this approach is beginning to be taken seriously by some scientists and engineers.

THE PRESENCE OF WOMEN
IN THE SCIENCES

The physical sciences still include an extremely small number of women, for the reasons we have discussed in earlier chapters, as summarized by Hilary Rose (1994, pp. 100–01):

> It was with the professionalization and industrialization of the sciences, and the steady transfer of the production of scientific knowledge from within to outside the home—first chemistry in 19th-century Germany, later physics and most recently biology—that women came to be systematically and sequentially excluded from the new occupational structures, which, at their apex, were linked to new forms of economic and social power. . . . This was to become an intensely meritocratic means of achieving and confirming class power for men, . . . and for very few women . . . who became increasingly confined to areas of more contemplative, less interventive science (such as botany in the 19th century) or to newly developed areas (biochemistry in the 1930s, crystallography in the 1940s and 1950s, computing in the 1960s and 1970s). These were areas with less fully elaborated career structures. . . . New industrialised science resembles a factory, a process of intensely skilled, closely focused teamwork with marked division of labour and hierarchy; then replaced by smart technology, deskilling, and robotization of scientific production.

It is clear that the notion of balance between nature and human beings, and between women and men, propounded by Paracelsus and others (see Chapter 1) has long been submerged because of the triumph of the concept of mastery of nature associated with, but of course by no means limited to, Sir Francis Bacon (see Chapter 4). Vandana Shiva (1989, p. 20) contrasts the two approaches:

> The mechanical school represented by Bacon created dichotomies between culture and nature, mind and matter, and male and female, and devised a conceptual strategy for the former to dominate over the latter. The two visions of science were also two visions of nature, power, and gender relations. For Paracelsus, the male did not dominate over the female: the two complemented each other, and knowledge and power did not arise from dominating over nature but from cohabiting with the elements, which were themselves interconnected to form a living organism.

Today, with the trend toward "big science" and very costly installations, women have found niches for their work in certain branches of science and in mathematics. Research work on mathematics or statistics, like the writing of novels, can be done in an office or a study, with little equipment apart from a computer. The craft mode of production in specific cultural activities, then as now, makes them more accessible to socially privileged women. This perhaps explains why there are more women in mathematics than in physics (Rose 1994, p. 16). Similarly, women students are present in almost equal numbers with men in biomedical engineering courses, which can lead to careers in research, teaching, or industry. Much of the work in this field can be done in a reasonably small clean space, with a computer and small-scale equipment for collecting and analyzing biological signals, and it rarely requires large investments, except, for example, in the case of work involving magnetic resonance imaging. Biomedical engineering is a blend of life sciences and engineering, and it is noticeable that biology at the undergraduate level tends to attract substantially higher proportions of women than of men.

There are, however, countries where, as Beatriz Ruivo (1987) has shown, "a large proportion of elite women are involved in semi-industrialized contexts because scientific production is still part of cultural production, and not yet fully locked into technological and economic growth." These included Portugal, Argentina, the Philippines, Mexico, and Turkey when Ruivo was writing, and today would include several other countries in Latin America and Asia. Historically, the participation of relatively large numbers of women in science and engineering also reflected deliberate policy choices in, for example, the former Soviet Union and its satellites: women formed half of the research staff at the Ioffe Physical–Technical Institute in Leningrad (formerly, and now, St. Petersburg) and over half of the research positions in the leading Polish biological institute, the Nencki, were occupied by women (Rose 1994, p. 104). Yet in these countries, as elsewhere, women tended to have secondary or subordinate positions, and their careers could be interrupted or ended because of marriage or childbirth.

SCIENCE AND THE PUBLIC SPHERE

The current rise of conservative values in many western nations has resulted in severe cuts in support for groups working on improving the status of women, and a concurrent decline in the proportions of women in science and engineering. In Russia and most of eastern Europe, for example, the new enrolment of women in these fields has declined since the 1990s. There is also a resurgence of religious values in many countries, and political and academic leaders are urging women to return home, develop their spirituality, and nurture the moral renaissance of their countries (Rose 1994, p. 105).

One possible deterrent for women considering careers in science or engineering is that, as David Dickson (1984) puts it (p. 107), there are "almost as many American [and other] scientists and engineers helping, directly or indirectly, to develop new ways of destroying life as there are trying to improve it." Dickson also contends (p. 83) that until recently it was superior force, military or civilian, that allowed particular individuals, groups, or social classes to control and have power over others, but since 1945 technology has become the key to both economic and military power.

Another current issue concerns the overwhelming pressure from corporations, banks, and the military to concentrate funding and personnel on certain profitable or politically prestigious activities, while neglecting research and development that may be much more socially beneficial. As the worlds of academia and finance converge, there is increasing potential for conflict between scholarly responsibilities and commercial demands, which can have an impact on what projects get done and on their scope. Moreover, the commercial ventures or spin-offs created by academics also create potential conflicts for scholars between the demands of their various roles and the amounts of time and effort they devote to each of them (Dickson 1984, pp. 81–82).

IS A GENDER-FREE CULTURE POSSIBLE IN SCIENCE AND ENGINEERING?

Since the women's movement began to make an impact in the 1970s, many books and articles have been written on feminism and science. For example, Evelyn Fox Keller (1986) explores the origins of modern science and discusses gender and science, as well as theory, practice, and ideology in the making of science. In her concluding chapter, she makes the point that not all men embrace a conception of masculinity that demands cool detachment and domination. She defines the aim of her essays as follows (p. 178):

> It is the reclamation, from within science, of science as a human instead of a masculine project, and the renunciation of the division of emotional and intellectual labour that maintains science as a male preserve. . . . [The essays] are preoccupied with androcentric bias in prevailing definitions of science; their aim throughout is the transcendence of that bias.

In the same spirit, Sandra Harding (1986) examined the trends in the feminist critiques of science, identified gaps, tensions, and conflicts between them, and concluded (p. 10) that these critiques can have "implications at least as revolutionary for modern western culture self-images as feminist critiques in the humanities and social sciences have had."

Evelyn Fox Keller later collaborated with Helen Longino (1996) on a collection of articles by several authors on representations of sex and gender, language, gender and science, and gender and knowledge. For example, Emily Martin (in Keller and Longino 1996, pp. 103–20) describes how preconceived notions of an active sperm and a passive egg, corresponding to stereotypical gender roles, distorted knowledge of fertilization, in which the egg and the sperm are actually mutually active partners—an excellent example of how preconceptions and stereotypes can produce bad science.

To take just one more example, Londa Schiebinger (1999, pp. 126–32) describes how Jane Goodall and Dian Fossey introduced new methods of observation of chimpanzees and silver-backed gorillas, and how this enabled scientists to record different perspectives on the behaviour of these primates, especially the behaviour of females; she also describes the contribution of other women, such as Thelma Rowell and Shirley Strum, who discovered more about the important roles played by females in nature:

> In primatology, as in medicine, the majority of feminist changes to date have come from re-evaluations of females. Only in the 1960s did primatologists begin looking seriously at what females do. Feminists first overturned the conventional stereotype of the passive, dependent female. . . . Primatologists questioned the stereotypes of male aggression, dominance and alliance, and female compliance. They studied the significance of female bonding through matrilineal networks, and analyzed female sexual assertiveness, female social

strategies, female cognitive skills, and female competition for reproductive success.

These and other feminist critiques have helped to identify issues in contemporary science that can lead to distorted knowledge, and even to errors in results and the inferences made from them. However, in my experience, most scientists and engineers today still believe in the objectivity of their work and fail to understand that bias can arise in many forms: in the research question and the manner in which it is posed; in the choice of subjects to be studied; and in the analysis, inferences, and conclusions drawn from the study. Science would be greatly improved by an increased awareness of these issues, as Evelyn Fox Keller suggests in her account (Keller 1986, p. 178) of what gender-free science would be like—it is not to be understood as

> a juxtaposition or complementarity of male and female perspectives, nor is it the substitution of one form of parochiality for another. Rather, it is premised on a transformation of the very categories of male and female, and correspondingly, of mind and nature. . . . A healthy science is one that allows for the productive survival of diverse conceptions of mind and nature. . . . it is not the taming of nature that is sought, but the taming of hegemony . . . The survival of productive difference in science requires that we put all claims for intellectual hegemony in their proper place—that we understand that such claims are, by their very nature, political rather than scientific.

Adopting such an approach would, of course, lead to increased awareness of potential biases (political, social, economic) in the development of science, and, in my view, lead to better science.

However, the majority of scientists and engineers, both men and women, are not aware of feminist discussions and critiques of science, and so the approach suggested by Keller is not likely to become part of their culture for many decades to come. The solution I propose to make science and engineering more gender inclusive is based largely on the suggestions made by Linda Jean Shepherd in her book *Lifting the Veil: The Feminine Face of Science* (1993), and on my own belief that the differences between men and women, arising from the different socialization of girls and boys— from nurture rather than nature—can be put to the service of better science.

Shepherd believes that blending feminine approaches and attributes with the current masculine culture of science and engineering could make positive contributions to the work accomplished in these fields. Mary Belenky, Blythe Clinchy, Nancy Goldberger, and Jill Tarule, whose book *Women's Ways of Knowing* (1986) is another good source of information on feminine approaches to learning, provide strategies for educators to help women in developing their selves, their voices, and their minds, arguing (p. 229) that

> educators can help women develop their own authentic voices if they emphasize connection over separation, understanding and acceptance over assessment, and collaboration over debate; . . . [and] if they encourage students to evolve their own patterns of work based on the problems they are pursuing. These are the lessons we have learned in listening to women's voices.

Women may be able to take a new approach in their work, but will they have an impact on what technologies are developed? It is likely that this will happen when more women have been promoted to decision-making roles. An alternative is for women to create their own technology businesses. Then they may decide what they want to develop, how these technologies will look, and how people will use them.

It is a fact, nonetheless, that not all women feel comfortable in using feminine attributes in male-dominated fields. According to Linda Jean Shepherd (1993, p. xiv),

> When we are enmeshed in a hierarchal framework, we automatically rank one person, profession, race, or gender over another. In doing so, we fail to value the wonder, beauty, and benefits of diversity. . . . As long as this hierarchal worldview predominates, being different from the white male professional means being inferior. . . . Many women, even feminists, are nervous about identifying with anything feminine, since they have worked so hard to prove their equality with men. As a result, they deny the parts of themselves that are different from men and are reluctant to explore any quality that could be labelled inferior, such as feeling or nurturing. Even the notion that women could be different from men is threatening.

As a young engineer in the 1960s, I tried very hard to be "one of the guys" and did not want to be identified as a "woman engineer." Admitting that you were different was synonymous with admitting that you did not belong. At the first conference I attended, in the United States in 1971, I was one of two women among one thousand men. I dressed conservatively, wore little jewellery, and asked serious questions, intended to be noticed for their cleverness, to help me make my way forward in the field. With hindsight, I can see that perhaps being a woman did make it easier for people to remember me, and that may have been the one advantage of my situation, but it took me more than fifteen years after graduation to think about what it meant to be a woman engineer, and to ponder why there were so few women in my field. This is a common occurrence among young women in any male-dominated field, and this period of denial makes it much harder to make progress on gender issues.

POTENTIAL BENEFITS OF INTEGRATING FEMININE PERSPECTIVES

Some interesting studies have addressed the question of whether women can bring to science and engineering perspectives that differ from men's in certain ways. The study by Anne Van Beers (1996), cited in previous chapters, found that several participants agreed that women could bring changes to the structure of the work environment, the culture, and the practice of engineering, and introduce values and perspectives that could be complementary to men's. One example is the impact that a woman civil engineer's experience as a mother had on a new idea in designing a building. In New Brunswick in the 1980s, a new terminal building for the ferry to Prince Edward Island included baby-changing tables in the washrooms for both sexes, which greatly facilitated the changing of diapers. The designer explained that her experience as a mother of two had helped to think of this feature. These tables can be now found in airports, train stations, and other buildings all over the world.

One feminine attribute frequently mentioned in the literature is that women tend to use a contextual approach, which means taking into consideration all the aspects of a problem. This usually takes a little longer than using the "bottom line" approach preferred by most men, but it can provide more complete solutions. It is also common knowledge that women tend to have well-developed communication and people skills, and that many women prefer more consensual working relationships and fewer hierarchical organizational structures. Women's affinity for consultative styles of working is very much in tune with today's management philosophy. The combination of verbal and interpersonal skills with a solid technical education can become a real asset, especially for smaller firms whose engineers must interact with suppliers, clients, and regulatory agencies. Similarly, individuals from other underrepresented groups who have been raised with different cultural influences may bring different and innovative solutions to engineering problems. Everyone benefits when diverse, gender-balanced teams design new products or solve environmental and technological problems.

Everyone can also benefit when women scientists and engineers draw attention to discrimination in a particular field of science or engineering. This has been true of medical research, for example, as Lesley Doyal (2001) points out:

> There is still evidence that women are treated by some doctors as less valuable than men. This can lead to demeaning attitudes as well as the unequal allocation of clinical resources. This gender bias is especially evident in the context of medical research, where women have often been excluded from studies for inappropriate reasons. . . . If the gender bias in medical research is to be eliminated, measures will need to be taken to ensure that study designs include sex and gender as key variables whenever appropriate. In the short term this would promote equity through filling the gaps that currently exist in our knowledge of women's health. In the longer term it would improve the overall quality of medical science and would therefore benefit men too.

In medicine, as in other sciences, it is clear that including both sexes in studies, and looking through a gender lens to see if the study addresses the needs of both women and men, would produce knowledge that can be more generally useful to society than just to half (or, in fact, in many countries, less than half) of the population.

In addition to gender attributes, paying attention to personality types can also bring about differences in approach. Drawing on the idea of Carl Jung, Katherine C. Briggs and Isabel Briggs Myers organized personalities into a system of sixteen types based on eight paired aspects of personality:

introversion and extraversion, sensing (S) and intuition (N), thinking (T) and feeling (F), and judging (J) and perceiving (P). Bart Noordam and Patricia Gosling (2007) provide a short description of how this system can help in understanding how decisions are made in science and engineering:

> Thinkers (Ts) are more likely to choose or make decisions based on impersonal information. In contrast, Feelers (Fs) make decisions that tend to be subjective, based on their value system, and take account of their decision's impact on others. Science is based on facts, so science and science-related fields probably attract more Thinkers (Ts). However, there may be a number of Feelers in your lab, and they can play an important role in thinker-dominated teams. Because Thinkers make decisions based on logic and reason, the input of Feelers can round out the process. There are some instances, even in research, where gut feeling and intuition have value. So if you're a thinker who prefers a rational, logical approach, try to be open to—and not frustrated by—the more intuitive side to decision making when confronted with a Feeler.

It seems evident that scientific work can best be accomplished by a group including both women and men, and a variety of personality types: some Ts and Fs, some Ns and Ss, some Js and Ps.

Linda Jean Shepherd (1993, p. 58) also describes the difference between Thinking types and Feeling types regarding their attitude towards science, suggesting that the Thinking type would be doing science for the sake of science, whereas the Feeling type would ask several questions: "Who sets the priority of projects and the consequence of knowledge acquired? Who will have access to the knowledge? Who decides what funds will be diverted from one project to another?" If more scientists were Feeling types, then perhaps these questions would be seriously debated and taken into consideration by funding agencies, foundations, and researchers themselves. Should spending on military research continue to be substantial, or even increased, or should more research go into reducing poverty, increasing literacy, and improving health around the world? Should we try to attract more Feeling types into science and engineering, to help in assessing their impact on society and to examine the ethical issues related to them? In my view, we should, but it is unlikely that these Feeling types would remain in these fields if the culture does not change, so that they are given opportunities to make their voices heard, and are respected for the challenges their perspectives bring.

However, since the earliest stages of the development of science, emotion and feeling have been seen as suspect, even though they can have a very positive impact on what science gets done and why. As Shepherd writes (1993, p. 58),

> Feeling types have a strong sense of values, reacting spontaneously to people, events, and ideas. Feeling judgments have their own rationality based on a sense of good and bad, right and wrong, beautiful and ugly, and levels of importance and harmony. Such judgments depend on the context of the situation, rather than on a prescribed set of rules. As such, they can provide the basis for evaluating priorities and ethics in science, which has long-laid claimed to being "value-free."

My own graduate course in engineering at Carleton University and the University of Ottawa includes a discussion on ethics and on the impact of science and engineering on society. At the beginning of the term, students mention that they have never given much thought to this matter. In this course, students have to write an essay describing one particular technology or science and discussing its positive and negative effects. Many of them change their attitudes by the end of the course and begin to assess their thesis projects with regard to their social impact and the ethical issues that may need to be considered. In short, they add a Feeling function to their Thinking one, showing that people can change their personality types over time as they gain experience and knowledge. The ideal in science, then, would be to have Thinking types who learn to add Feelings to their work, so that they can use both logic and values in making decisions.

This also raises the question of subjectivity in research. In my view, subjectivity is not a necessary or inherent feminine characteristic, but perhaps some women are more aware of its relevance in science and engineering than the majority of scientists and engineers, who still believe that science is, or at least should be, value-free and objective.

Being involved in science in the 17th century, and even in the 18th century, required tools and free time, so this activity was mainly done by members of well-to-do classes. In the 20th and 21st centuries, one needs funding to buy equipment, to pay student assistants, and to cover publication costs. The foundations and agencies that fund research often have priorities, for topics and projects that will be funded. Other sources of funds are the military and industry, which have their own priorities and agendas. All the funding decisions thus involve some kind of politics, and this is certainly not value-free. Industry-supported research may also lead to the suppression of publication for the sake of applying for patents, or to keep information secret for reasons of "commercial confidentiality." Even universities now encourage the protection of the "intellectual property" of their researchers if there is an opportunity for technology transfer into the marketplace. As Linda Jean Shepherd succinctly puts it (1993, p. 102), "Throughout history, science has often served the builders of empires—financial and political."

Thus, in spite of claims about "value-free" science, it is necessary to concede that the reality is different, and that political, financial, and social considerations are often interlaced in science and engineering projects, potentially creating bias. In these circumstances, can subjectivity become an asset for research? I would argue that if we wish to minimize potential biases when identifying the objectives of research, the methodology, the subjects to be included, and the variables to be measured, it is indeed important to understand how subjectivity can have an impact on our work.

Another attribute often dismissed by scientists is receptivity. This means listening patiently to data and observations, as, for example, Jane Goodall and Dian Fossey did. It means reflecting on what the data and observations are showing, instead of fitting the data to a preconceived model. Linda Jean Shepherd (1003, p. 83) discusses the approach of the biophysicist Cynthia Haggerty, who studies the pathologies of fish with an electronic microscope, constantly asking herself, "What is there for me to see? What is there for me to know? What does this material *tell* me about the life process and about whatever pathology is going on?" This open-minded, receptive attitude can lead to new knowledge, while trying to fit the data to a preconceived model—an approach sometimes referred to as "data torture"—can lead to serious errors, or to missing key points.

Finally, if the culture of science and engineering, and with it the results of their activities, are to be improved in the ways discussed in this section, the teaching of these subjects will also have to change. Linda Jean Shepherd (1993, pp. 159–60) advises teachers

> to give students a chance to develop independence while providing gentle guidance. The idea is not to stifle innovation and creativity, but to nurture it through mentoring and support, while encouraging exploration. . . . A more nurturing attitude in science allows us to expose our ignorance without fear.

Several universities have, for example, instituted special measures to retain first-year students. Others have special tutoring for students who have difficulties in one particular subject. In my own laboratory, postdoctoral students mentor doctoral students, the latter mentor students at the master's level, and fourth-year undergraduate students involved in senior thesis work are mentored by graduate students to ensure that they perform well on research projects that develop their interest in graduate studies. The work of these undergraduate students has even led to publication in certain cases and has served them well in their graduate work. All students and researchers— physicians, engineers, and cognitive and computer scientists— meet as a team on a regular basis and everyone contributes ideas for ongoing projects. Everyone is treated as an equal and an important member of the team. This nurturing environment has been very successful not only in retaining students but also in encouraging them to embark on doctoral studies. All have defended their theses well and obtained meaningful employment after completing their degrees.

FOSTERING COOPERATION

Science and engineering are fields where competition is fierce. There is competition for grants, for recognition, and for visibility at conferences and events. This is all part of the culture described in previous chapters. However, it is possible in one's own laboratory or other workplace to foster collaboration among students and colleagues with whom one is sharing a grant or co-supervision of theses. The workplace itself, of course, needs to get funding and to strive for greater visibility, but within its walls students or colleagues can learn to help and support each other and to share information that helps the whole group to move forward. Secrecy may help one particular individual, but likely at the expense of the entire group. Open, transparent meetings where everyone feels free to bring up questions and comments, and ask for help or offer suggestions to others, even to the head of the group, creates a positive environment and enhances trust.

I have frequently observed that women researchers' laboratories function successfully in this manner. Students are respected and develop self-confidence and assurance. They are exposed to broader knowledge by the sharing of progress reports from all the students working on a variety of projects. They do test runs with each other before making presentations at major events and conferences. All this fosters excellence for all students and colleagues.

Linda Jean Shepherd rightly deplores the fact that, in contrast, most research locations are organized in hierarchical structures (1993, p. 139):

> In the pyramid of hierarchy, someone at the top must be displaced to make room for another person eagerly climbing to the summit. In a circular structure, people meet at eye level and everyone inhabits the same level. The circle can expand to include others without displacing someone. But the circle has only one level, and so it can foster sameness and repetition, and has the disadvantage of hindering advancement of individuals. . . . The spiral, on the other hand, embraces both multiplicity and advancement. Any level can expand to include another person as each individual grows. The head of the lab may be at another level, but in a spiral each level is continuous with all other levels. No one needs to be displaced.

In a hierarchical, competitive environment, researchers keen to get their results published as early as possible may distort or invent data. Examples of fraudulent research are numerous, most prominently in the field of medicine, which the media pay more attention to than they do to other disciplines. In Canada, for example, Dr. Ranjit Chandra, who for many years was a researcher at Memorial University of Newfoundland, claimed that when his elderly test subjects took his combination of vitamins and minerals, their brain functions, including memory, showed dramatic improvement.

The *British Medical Journal* refused to publish Chandra's results and another journal that did publish some of his research issued a retraction. Chandra eventually resigned from Memorial University, but now sells his preparations through a mail-order company. In the United States between 1987 and 2001, Dr. Eric Poehlman, who then held posts at the Universities of Vermont and Maryland, published more than two hundred journal articles and received 2.9 million US dollars in research funding from the National Institutes of Health, the Department of Agriculture, and the Department of Defense. In 2005, he resigned from the Université de Montréal, which had appointed him on the basis of his apparently impressive research record, and admitted altering and fabricating the data included in seventeen of his grant applications and ten of his journal articles. He was barred for life from seeking federal funding, had to retract or correct journal articles, and pleaded guilty to one count of making a false statement in a federal grant application, for which he was ordered to serve a year and a day in a federal prison (see Goldberg and Allen 2005).

The "publish or perish" culture, which contributes to the climate in which committing fraud can seem tempting, needs to be replaced by assessment of the quality of scholarly work rather than merely its quantity. Criteria need to be revised to effect this change of culture. Some university deans and presidents have suggested to me that one way to do this might be to have the assessment focus on the best three papers or the best five papers a scholar has produced, depending on the recognition level being considered. How maternity or paternity leaves are counted is also of great importance. Some granting councils and universities do not count these career breaks when measuring productivity unless the person in question asks for them to be included. This helps in measuring productivity when it should count, that is, when the person is fully active.

The culture in some of the funding councils is in fact beginning to change. For example, in Canada the Natural

Sciences and Engineering Research Council (NSERC) has put in place maternity/paternity policies that allow researchers to use their grants for baby-care services while attending conferences. An attempt has also been made to appoint more women to grant selection committees and to the NSERC itself, and to allow postdoctoral student applicants to remain at the same institution if they are not able to move because of a spouse's or partner's employment in the same city. Grant applications now contain a section in which researchers can explain delays in publication. Moreover, the NSERC has created special grants for women and Aboriginal applicants who meet its excellence criteria. These grants while they existed provided salaries and research funding for a period of five years.

One additional problem is that, in any funding system, it is difficult—in fact, nearly impossible—to win a grant to work on new ideas or projects. Acceptance of new data or new theories often depends on whether they fit into the prevailing view of how the world works, and it is noteworthy that it is easier to pass off fraudulent work if it appears to agree with widely believed theories (see Shepherd 1993, p. 88). Funding bodies tend to expect efforts made in doctoral work to be continued by young assistant professors in their first grant applications, and future requests for funding are usually expected to show some continuity with established research. In this sense, little has changed since 1982, when, as reported by Linda Jean Shepherd (1993, p. 101),

> in a . . . survey of grant applicants to the [US] National Cancer Institute, 60 percent of scientists held that reviewers are reluctant to support unorthodox or high-risk research; only 18 percent disagreed with that assertion. They said: Anything novel had to be bootlegged. One must never say anything new in a grant application. The proposals that get funded are generally the most boring and mundane.

It is possible to make a transition to new work, but it must be carefully planned. For example, one can publish on the main research project for which funds have been granted, but also publish papers in a new area or on a new project. This may help to bridge future grant applications to the new project and allow the researcher to slowly detach from older ones. I have done this by publishing in two distinct areas of research, clinical decision support systems and infra-red medical imaging. Now that both areas are well established in the laboratory, grant applications can be requested in either of them.

Some funding agencies are also now taking steps to have applicants include in their grant proposals some account of the potential benefits to society from the research they are proposing to do. In 1997, the National Science Foundation in the United States, which had traditionally sought to judge proposals on the criterion of intellectual merit, added a second criterion, asking researchers to discuss the broader impacts of their proposed activity (Holbrook 2005, p. 439). Some applicants have resisted completing the section of the application form dealing with this second criterion, but over time it is likely to become more broadly accepted. Similarly, the Canadian Foundation for Innovation, in its first year of operation, 1997, required that applicants discuss the "benefits to Canada" of their proposals. These are questions that should make researchers think seriously about the relationship between science and society.

PART IV

Profiles of Three Women

by Peter Frize

—◦◦◦—

The Bold and the Brave: Sophie Germain, Mileva Marić Einstein, and Rosalind Franklin

Throughout the brief history of human beings, men have been recognized for their achievements in all spheres of endeavour. Their ubiquitous presence has been observed and documented in many forms of literary production. In government, the military, religion, and commerce, they have been prominent in the upper echelons of power and control. In our early scholastic years, we are made aware of their achievements, past and present, and of their contributions to the progress of civilization. "Lest we forget," statues have been erected in their honour, buildings, bridges, roads, airports, and highways have been named after them, and paintings have been commissioned to ensure their posterity. Young men with high aspirations and ambition certainly have a plethora of role models to choose from and emulate.

For young women, especially in science and engineering, it is a different story. They know of few female role models to admire and emulate. This is not to suggest that such role models do not exist. On the contrary, there are many, from the past to the present, who, unlike their male counterparts, have been either ignored or rendered invisible by historians. If students in a science class were asked to name as many female scientists as they could, most would be hard pressed to proceed further than Marie Curie, and they would probably be unaware that her daughter, Irene Joliot-Curie, also won the Nobel Prize for Chemistry, in 1935. Yet the list of brilliant women is long, even if our collective memory of them is short. The purpose of this chapter is to shed some light on the lives and work of just a few of the bold and brave women who have challenged the male status quo, often to the detriment of their health, even of their lives.

For men, the world, and even the universe, have been available for their personal discovery, influence, and profit. The knowledge they accumulated, coupled with the assets they acquired, were passed on to the next generation of male progeny. The continuous flow of information and assets only to male offspring left more than 50 percent of the world's population at a great disadvantage, unable to achieve financial independence. Historically, a woman's role was defined and strictly adhered to. Her world was relegated to a small area, and her *raison d'être* was to procreate. The only information deemed necessary for her was about domesticity, and the only knowledge that she acquired was that her lowly position in life was preordained from cradle to grave.

In today's fast-paced world, which is inundated with electronic gadgets and information overload, it can be difficult to pause and take stock about how we should live. Many people no longer have time to reflect, only react. How we use the technology and information that is so readily at hand can have a positive or a disastrous effect on humanity. For example, when the ultrasound monitor came on the market in the 1970s, its purpose was to monitor the health and status of the foetus, but unfortunately it could also determine its sex. In several developing countries, opportunistic clinics have opened up on street corners, advertising prenatal information on the sex of the foetus. In China, India, and other countries where female children are looked upon as a burden and not as a blessing, the acquisition of this information is frequently followed by an abortion if the foetus is female. Sadly, this demonstrates a cultural attitude in some countries that prizes the male child, resulting in millions of women missing from the planet.

When I agreed to collaborate on this book, my task was to discover and profile a few women who had bravely and boldly involved themselves in the sciences at times when this was exceptional. Before I carried out this task, my own awareness of women in the sciences was minimal. During my early schooling, the achievements of women were rarely mentioned in history lectures: "her-story" was submerged by "his-story." However, although my own discipline is in arts and not in science, I do have the benefit of being married to a scientist (an engineer), whose guidance and patience were essential. My reading material was now redirected from novels into the world of female scientists. Over the past twenty years, there has been a flurry of activity and interest among many distinguished authors who have decided to reincarnate the forgotten and invisible women from the past. The emergence of women's studies from the late 1960s has been a major factor in stimulating research and furthering our enlightenment about so many women. "Her-story" flooded my senses. I felt like the prisoner in Plato's allegory of the cave, released from vague shadows into the light of reality. My reading produced a whole gamut of emotions that oscillated from awe and delight to anger and sadness. Questions churned within me. Why have so many historians either ignored or rendered women invisible? Why is this crime against the female intellect still being committed?

If I had to choose a common denominator among all the women who have entered the scientific field, it would be their passion for understanding and discovering the natural world around them. (This should come as no surprise: if one wishes to refer to our genesis, we should take note that it was Eve who desired knowledge. Could this indicate that women, by their very nature, are more curious than men?) The inquisitive nature of women makes them ideal candidates for scientific research. In her book *The Scientific Lady* (1990), Patricia Phillips mentions that in the mid-17th century, gentlemen regarded scientific pursuits as a "common" and unnecessary undertaking. Only the reading of the Greek and Latin classics was worthy of their attention. She points out, wryly (1990, p. 8), that "the classics marked, as nothing else could, not so much the scholar as the gentleman."

Throughout this book there have been small cameos of many bold and brave women, role models for readers to admire and, if they wish, to emulate. In addition to their lives, we now present more detailed profiles of three women who were not necessarily privileged, but were determined to make a difference in their own lives and the lives of others. They were bold, and they bravely challenged the status quo on women's roles.

SOPHIE GERMAIN (1776–1831)

Sophie Germain is a prime example of an eminent female mathematician whose name remained, until relatively recently, conspicuously absent from history books. Only the admiration and respect she received from various male mentors and colleagues saved her achievements from being completely lost forever.

Her most important contributions to mathematics were her work on Fermat's last theorem and on the theory of elasticity. Regarding Fermat's last theorem, she showed that for all prime numbers n less than 100, there are no solutions for the case in which one of the three numbers x, y, and z is divisible by n (Bucciarelli and Dworsky 1980, p. 86). Fermat's last theorem, $x^n + y^n = z^n$, has no positive integer solutions for x, y, and z when n is greater than two. As for her work on the theory of elasticity, which was submitted for a prize offered by the

Emperor Napoleon I, Merry Maisel, and Laura Smart (1997) comment that

> while it contained mathematical flaws and was rejected, her approach was correct. All the other possible entrants in the contest were prisoners of the ruling paradigm, consideration of the underlying molecular structure theorized for materials. The mathematical methodologies appropriate to the molecular view could not cope with the problem. But Germain was not so encumbered.

Sophie Germain was born on April 1, 1776, into a bourgeois family in Paris. She was the middle child, between two sisters, both of whom married professional men. Sophie, however, was shy and reclusive, and she chose to remain single and to commit her life to mathematics. Her parents felt that mathematics was an inappropriate undertaking for a young lady, but her dedication was so great that their attempts at dissuasion, such as keeping her room cold and confiscating her candles, were to no avail. She simply wrapped herself in more blankets and retrieved her hidden cache of candles.

It has been suggested that her fascination with mathematics originated after she read the legend that Archimedes had become so engrossed in his mathematical calculations that he failed to respond to the intrusion of an enemy soldier, who promptly dispatched him with a sword. Sophie may have surmised that mathematics must be extraordinarily fascinating if someone could be absorbed in it to the point of being completely unaware of his imminent demise.

Sophie's father, Ambroise-François, became quite wealthy in the silk trade, and later obtained a directorship at the Bank of France. He came from several generations of business owners, some of whom were liberally minded and attracted to the educated bourgeoisie. Their political thinking was probably influenced by those authors of the 18th century whose attitude to the French Revolution was more radical. The main intent of the bourgeois-led revolution, which began when Sophie was only thirteen years old, was to rid France of the monarchy and the nobility and transfer ownership of land and property to the people. Sophie often listened intently when her father invited radical thinkers into his home to discuss revolutionary matters and social change. Her exposure to these discussions probably influenced her own independence of thought and behaviour when, later, she was faced with the biases and prejudices of the elite in her own particular field of endeavour. Her intellectual curiosity was further stimulated by her daily visits to her father's well-stocked library. She spent many hours there, alone and unsupervised, studying Greek, Latin, and mathematics, while the bloody turmoil of the French Revolution was played out in the streets of Paris.

Sophie was nineteen years old when the École Polytechnique was established in 1795. The founders of this military engineering academy made the unfortunate decision that no women would be admitted as students, a patriarchal prohibition that proved to be highly detrimental to Sophie's career, imposing difficulties upon her that her secure and well-educated male counterparts rarely had to confront, as Margaret Alic (1986, p. 155) explains:

> Lacking both formal training in mathematics, and access to the most recent publications and the latest thinking in her chosen field, Sophie was writing *memoirs* [articles] that were clearly inadequate and could not be approved by the academy. . . . . But the work was all being done within a community that excluded her so completely that she did not even realize what was happening. It was her sex, not her mathematical ability, that was the determining factor.

Although the École Polytechnique refused admission to women, it did provide lecture notes to anyone who requested them. Sophie duly requested them, and read them, but she soon realized that lecture notes were insufficient if she wanted to make further progress. With a touch of bravado, she decided to use the pseudonym of a student she had befriended, a Monsieur Leblanc (or, in some sources, Le Blanc), who had recently left Paris. The administration of the École Polytechnique, unaware that he had left, continued to send him lecture notes and problems to solve. Sophie intercepted the material and forwarded her own answers to the professors. Knowing that they would be rejected if the professors knew that she was a woman, she submitted several papers to Professor Joseph Lagrange using Leblanc's name. Impressed by the quality of the papers, and surprised at the sudden improvement in the work of a student who previously had been mediocre, Lagrange requested an interview with him. Instead of being angry and upset at being deceived, he was delighted to discover that the talented student was a woman. With her newly found mentor, Sophie now felt more secure, and could continue her studies with confidence and without the necessity of further deceptive tactics.

For Sophie, the assistance of Professor Lagrange opened doors that had previously been closed. She had now gained acceptance, albeit by correspondence, into the inner circle of the academic elite. Sophie's confidence in her mathematical ability soon began to expand. She now became deeply involved in number theory, and this led her to discover Pierre de Fermat's last theorem.

The leading expert on number theory at this time was the German mathematician Johann Carl Friedrich Gauss (1777–1855), whose magnum opus *Disquisitiones Arithmeticae* (written in 1798 and published in 1801) presented number theory as a solid foundation for all areas of mathematics. Still fearing rejection if her gender was revealed, Sophie corresponded with Gauss using the identity of Leblanc. At that stage, Gauss had little interest in producing a proof of Fermat's last theorem, regarding the problem as a waste of his precious time. When a friend and later colleague of Gauss's, the professor and astronomer Heinrich Olbers (1758–1840), urged

Gauss to compete for a prize offered by the Académie des Sciences in Paris for a proof of the theorem, Gauss responded (as quoted in Singh 1997),

> I am very much obliged for your news concerning the Paris prize. But I confess that Fermat's last theorem, as an isolated proposition, has a very little interest for me, for I could easily lay down a multitude of such propositions, which one could neither prove or disprove.

However, his correspondence with Sophie Germain eventually sparked Gauss's own interest in Fermat's theorem, the proof of which had already been attempted by several mathematicians, including Euler and Legendre, though it was Sophie Germain who was one of the first to provide a partial solution for a large class of exponents.

Gauss's discovery of Sophie's gender came about by chance. In 1806, when Napoleon invaded Prussia, Sophie, fearing for Gauss's life, asked a friend, General Joseph-Marie Pernety, to guarantee his safety. He fulfilled her wish, but happened to mention to Gauss that his saviour was a Mademoiselle Germain. Sophie was informed of this exchange and believed that she could now inform Gauss of her true identity. Gauss responded with delight and admiration, and wrote (as quoted in Bucciarelli and Dworsky 1980, p. 25),

> But how can I describe my astonishment and admiration in seeing my esteemed correspondent, M. Leblanc, metamorphosed into this celebrated person, yielding a copy so brilliant it is hard to believe? . . . But when a woman, because of her sex, our customs and prejudices, encounters infinitely more obstacles than men in familiarizing herself with their knotty problems, yet overcomes these fetters and penetrates that which is most hidden, she doubtlessly has the most noble courage, extraordinary talent, and superior genius.

Unfortunately, their scholarly collaboration ended abruptly, for Gauss's interest in number theory diminished after his appointment as Professor of Astronomy at the University of Göttingen in 1808. Undeterred, Sophie now set out on a path that was to lay the foundation for a modern theory of elasticity.

In 1809, the Institut de France offered a medallion of gold weighing one kilogram to any person who could formulate a mathematical theory of elasticity. After the two years allotted for the contest had expired, it was found that Sophie's paper was the only one submitted. The mathematical background required to solve this problem was unavailable, not only to Sophie but to several other mathematicians at that time, and her first paper was rejected. As mentioned earlier, her exclusion from formal education, as well as her social status and her gender, continued to hamper her progress, as Louis Bucciarelli and Nancy Dworsky (1980, p. 40) explain,

> Before Sophie Germain could begin to "study these phenomena in a thousand ways" she had to find a solid framework from which to begin. Such a base is generally taken for granted by members of a profession; it consists of knowledge and methodology so well integrated into one's professional understanding that it can be relied upon without special attention. . . . Sophie Germain had no such well-formed working knowledge: her education had been too haphazard and her continuing study too limited and random.

Although there were many mathematical miscalculations in her first paper, her radical ideas and new approach stimulated other mathematicians to delve further into the problem. In spite of this initial setback, she was determined to succeed and plodded on. The deadline for the next submission was extended for two years. Germain submitted another paper in 1811, being once again the only entrant, but this too was deemed unsatisfactory, although, because of the level of difficulty of the problem and the satisfactory manner in which she linked her theoretical results with those from experiments, she received an honourable mention. The competition was extended for yet another year, and in 1816 she finally won the Prix Extraordinaire.

It is interesting to note that although Sophie Germain's gender placed her in the position of being an amateur, her first paper had stimulated Lagrange and others to explore the theory further. The jury at the Institut who awarded her the prize was comprised of the most eminent mathematicians of the time: Laplace, Lagrange, Legendre, Lacroix, Malus, Poisson, and Carnot. However, despite her stimulating work, Sophie remained a stranger to the scholarly community in her field and was excluded from all meetings, debates, and discussions. Inviting a single woman to a brain-storming session would have required a chaperon to accompany her to and from its location. Sophie was a very shy and reclusive person, and it was beyond her capabilities to request such assistance from a member of the opposite sex. Only the wives of the mathematicians attending were allowed to be present.

Thus, Sophie's training remained fractional and spotty. This had not been a problem for her work on number theory, because this form of mathematics was concentrated in one or two written works, and she also had the beneficial collaboration of several noted mathematicians, who encouraged her progress. However, for the mathematical physics involved in elasticity theory, her exclusion from the necessary expertise and information was a handicap. In spite of this, she made significant contributions to the field, and one may surmise how much more she would have achieved if she had the same support and recognition that Siméon-Denis Poisson (1781–1840), her chief rival and at times her main adversary, had received from his male colleagues.

Poisson and Germain took different stands on the theory of elasticity. The main difference between their approaches was that Poisson used the molecular theory, based on Newtonian physics, that was popular at the time, and that was based on the attraction and repulsion of molecules. His resulting equation was non-linear, but it was wrong. There was some incoherence in his results and his approach, but his work seemed well accepted by his colleagues. He refused to meet Germain as a fellow scholar, and ignored her in public. How much he may have influenced the prize jury one can only imagine.

When Germain received the prize from the Académie des Sciences for her third paper on elasticity theory, she chose not to attend the ceremony. It has been suggested that she was upset at the judges because she felt they did not appreciate her work and that the scholarly community did not show her the respect she felt she rightly deserved. She may have finally come to the conclusion that it was her gender and not her mathematical ability that had been the major obstacle on her chosen path. For her, pure mathematics was asexual and devoid of emotion. Poisson, in his attitude towards Germain, seemed confused by these two obvious traits.

Poisson, a student of Lagrange and Laplace, advanced rapidly at the École Polytechnique, met Laplace frequently for mathematical discussions and was soon elected to the Institut de France, while receiving several teaching positions. On August 14, 1814, he read to one of his classes a paper on elastic surfaces, presenting the material as his own and not divulging the fact that Sophie Germain had written it.

Another mathematician, Jean-Baptiste Fourier (1768–1830) also clashed with Poisson on several occasions. When Fourier became Secretary of the Institut de France, Sophie's fortunes changed for the better, for he invited her to the Institut for formal meetings, and her friendship with him gave her the feeling that she was participating in the scholarly community at last. She now undertook new studies on number theory with Legendre, on an equal footing. These important contributions still bear her name today.

In 1821, Sophie published the paper for which she had been awarded the prize, *Recherche sur la théorie des surfaces élastiques.* She wanted to ensure that the work would be properly attributed to her, for posterity's sake. In 1830, a few months before her death, she published her *Mémoire sur la courbure des surfaces.* Philosophy was another passion of Sophie's, and just before her death she wrote *Considérations générales sur l'état des sciences et des lettres.* This essay, published posthumously, was later praised by the philosopher and sociologist Auguste Comte (see O'Connor and Robertson 1996).

After Sophie Germain died in 1831, having suffered for two years with breast cancer, Gauss spent many months persuading and finally convinced the University of Göttingen to award her a posthumous honorary degree. The most noteworthy honour was given to her by H. J. Mozans in his pioneering study *Women in Science* (1913), when he wrote (see Mozans 1974):

> All things considered, she was probably the most profoundly intellectual woman that France has ever produced. And yet, strange as it may seem, when the state official came out to make her death certificate, he designated her as a *"rentière-annuitant"* (a single woman with no profession) and not as a *"mathématicienne."* Nor is this all. When the Eiffel Tower was erected, there were inscribed on this lofty structure the names of seventy-two *savants.* But one will not find in this list the name of that daughter of genius, whose researches contributed so much towards establishing the theory of the elasticity of metals, Sophie Germain. Was she excluded from this list for the same reason she was ineligible for membership in the French Academy, because she was a woman? If such, indeed, was the case, more is the shame for those who were responsible for such ingratitude towards one who had deserved so well of science, and who, by her achievements, had won an enviable place in the hall of fame.

MILEVA MARIĆ EINSTEIN (1875–1948)

One of Monique Frize's duties as the chair for Women in Science and Engineering in Ontario from 1997 to 2002 was to visit schools and talk to young women about considering careers in science and engineering. At the completion of her presentation, she would often leave them with this advice: "Choose your husband as carefully as you would your career." She realizes that this is far easier said than done.

Mileva Marić's first and probably only love affair was with Albert Einstein (1879–1955), whom she met in Zurich in 1896, when he was seventeen and she was twenty. Both were then students at the Eidgenössisches Polytechnikum, the prestigious Federal Polytechnic Institute (sometimes referred to in works on the Einsteins by its current name, the Eidgenössische Technische Hochschule, which it did not acquire until 1911). An attraction immediately developed between them, due in part to their shared passion for science and music. Unfortunately, this attraction was to lead to a marriage in which Mileva's life gradually deteriorated into desperate uncertainty, constant upheaval, and finally despair.

Mileva Marić was born on December 19, 1875, in Titel, a town that was then in southern Hungary but is now in northern Serbia. Its Serb majority population included Mileva and her family. Her parents, Milos and Marija Marić, were relatively wealthy, and had two other children, another girl, Zorka, and a boy, Milos. Mileva was a gifted child who expressed a passionate zeal for everything that intrigued her. Her eclectic interests included painting, handicrafts, music, medicine, psychology, mathematics, and science. In addition to her native language, Serbian (then usually known as Serbo-Croat), she achieved a high level of fluency in German, French, and Greek.

In 1882, Mileva, now seven, began her schooling. Even at this tender age she revealed a gift for mathematics and exhibited a photographic memory. In 1886, she entered high school in Novi Sad, another Serb-majority town also in Hungary but now in Serbia, with first-class credentials, but in 1890 she was transferred to the *Gymnasium* (secondary school) in Šabac, inside the then Kingdom of Serbia, to prepare for entry into higher education. For the next three years, her academic skills blossomed and she finished her courses there with a grade of First Class with Distinction. During her last year at the school, she was the only female student granted permission to study physics alongside the male students, and showed she belonged by achieving the highest grade ever awarded by the school in mathematics and physics.

In 1896, Mileva had to make a choice between studying physics or medicine. In the spring of that year, she took her matriculation examination at the Bern Medical School in Switzerland, and then transferred to the University of Zurich to continue her studies. For reasons unknown, however, she suddenly dropped medicine, and in the fall of 1897 she transferred to the Eidgenössisches Polytechnikum to study mathematics and physics. This would have been a remarkable and brave decision for Mileva to make, even more so if she were cognizant of the misogynistic atmosphere that prevailed in Swiss universities. Senta Troemel-Ploetz (1990, p. 424) wryly explains the patriarchal culture of Swiss universities, past and present:

> As for Swiss universities today, suffice it to say that in 1983, Switzerland could count forty women full professors, compared with over 2,000 men full professors, which is certainly an achievement in the 150 years since the first women audited courses at the university. At this rate, Swiss universities may actually reach 10 percent in another 600 years.

During Mileva's childhood, she had suffered from a hip deformity that caused one leg to become shorter than the other. She was also homely looking, and this, coupled with her prominent limp, often aroused ridicule and teasing from her schoolmates. Her unique academic skills also were not spared from their philistine attacks. Although this bullying made her more determined to succeed, it also left her with a poor self-image and an inability to fully appreciate her own qualities and talents. These two psychological drawbacks eventually became detrimental to her overall well-being as her life unfolded.

Albert Einstein, Mileva's future husband, had been born in the German city of Ulm on March 14, 1879, to Pauline and Herman Einstein. In 1891, they had a second child, a daughter, Maria, whom they affectionately called Maja. Apparently, the arrival of his sister was a disappointing event for young Albert. When his parents told him that he would have someone to play with, he imagined a new toy and it is said that, upon seeing his sister, he quizzically asked: "Yes, but where are its wheels?" (Highfield and Carter 1993, p. 12) During his early years he was prone to throwing temper tantrums and Maja felt the brunt of his anger: it is reported that she was often on the receiving end of various objects hurled her way. On Albert's fifth birthday, his parents hired a private tutor who quickly became a fresh target for his violent outbursts: "He also hit his private teacher with a stool—[and] frightened her so much that she ran away and was never seen again" (Highfield and Carter 1993, p. 12).

At the age of seven, Albert's parents enrolled him in a Catholic primary school. Being the only Jew in his class, he now found himself on the receiving end of frequent verbal and physical abuse from his anti-Semitic classmates. He was soon relegated to the margins of their group activities, including sports, which he disliked anyway. Fortunately, when Albert was five, his parents bought him a violin, which he took to quite eagerly. This instrument soon became his constant companion, providing solace in the solitude that was imposed upon him by his peers.

Einstein's biographers have offered differing opinions on Albert's early intellectual development. Ronald Clark, for example, claims that he lacked fluency in his speech, even by the age of nine, and responded to questions only after considerable reflection, causing his parents to wonder if he was more backward than other children his age (Clark 1971, p. 10). Conversely, Roger Highfield and Paul Carter are more positive about Albert's slow development, suggesting that what he later achieved gives hope to others who begin life with mediocre abilities, and that the prediction by Albert's mother Pauline that her son would be a great professor one day may indicate shrewdness on her part, rather than just optimism.

One area where Albert showed little backwardness was in his sexual development. Among physicists it has long been generally known that he was a womanizer, and committed adultery not only during his first marriage, to Mileva, but also during his second, to Elsa Löwenthal. Before he turned seventeen, he had a brief love affair with Marie Winterler, the daughter of a schoolmaster in whose home Albert boarded. After Albert had left for Zurich and met Mileva, he gradually ended his affair with Marie, although he "was still sending Marie his dirty laundry so she could wash it and mail it back to him" (Highfield and Carter 1993, p. 27). Albert had discovered at a very early age how useful women could be, both sexually and for domestic purposes.

Albert was a capable but difficult student who disliked and rebelled against the strict discipline of the 19th-century German educational system, as Carol C. Barnett (1998, p. 161) explains:

> Albert Einstein did not have as illustrious an academic background as Mileva Marić. This may have been in part due to his attitude towards structured education. . . . He preferred to pursue things that were of interest to him, and at times he was considered to be inattentive and disruptive by some of his professors.

Evan H. Walker points out that "by 1900 Einstein's grades were down. Albert passed with a questionable 4.91 average, trailing well behind his classmates Jacob Ehrat, Marcel Grossmann, and Louis Kollros"; he adds that "he barely satisfied the requirements for the degree" (Walker 1989, p. 122).

Putting more pressure on Albert were his problems with various professors, notably his thesis adviser, Heinrich Weber, as Andrea Gabor explains (1996, p. 13):

> Einstein eventually developed serious problems with Heinrich Weber, the professor of physics at the ETH [*sic*—it was then still called the Eidgenössisches Polytechnikum]; by the time he graduated, there was so much bad blood, in fact, that Einstein was convinced that the reason he had trouble finding a job after graduation was that Weber was sabotaging his efforts.

Weber was also Mileva's thesis adviser. Initially they had a positive working relationship, but on several occasions she clashed with him over his obvious dislike of Albert. When Weber refused to recommend Albert for a job upon graduation, Albert and Mileva came to the conclusion that there was no future for either of them with Weber.

Another major obstacle the couple had to overcome was the disapproval of their parents. Albert's parents objected to Mileva because, as Andrea Gabor explains, "she was not Jewish, because she was older than he, and because she was physically lame and an intellectual" (Gabor 1996, p. 11) The main objection from Mileva's parents was that Albert was Jewish.

In the summer of 1900, Mileva suffered her first academic setback by failing the final examination at the Eidgenössische Technische Hochschule. Andrea Gabor puts forward several plausible reasons for this failure, perhaps primarily that Mileva lived in fear of Albert's parents' tenacious disapproval of her. She was now determined that Albert should succeed and sought to prove to his mother Pauline that she was worthy to be his wife. She began to cling to Albert for security, gradually putting aside her own studies and sublimating her career in favour of his. According to Andrea Gabor, during the academic year 1899–1900, Mileva took on a more "wifely" role: if she did not cook, Albert would not eat. He now adopted a bohemian style of existence, often walking barefoot around the house, and his general appearance became more dishevelled. Friends stopped visiting them because they disliked the way Albert appeared to be exploiting Mileva (Gabor 1996, p. 12).

A crucial event that occurred in the early summer of 1901 has been mentioned by several biographers as marking the beginning of Mileva's decline and fall into obscurity. The conception and eventual birth of their daughter, Lieserl, could not have come at a worse time for Mileva. Gerald Holton agrees with several of Mileva's biographers that this illegitimate pregnancy was the beginning of her slide into despair (1994, p. 40): "There is no doubt that the birth of the real Lieserl took a heavy toll, above all on Mileva. More and more after these secret events, she tended to brood. Perhaps she was blaming herself for the decision to give up her daughter, or she blamed Albert for acquiescing in it." When Lieserl was born, secretly, in Serbia in early January 1902, a decision had to be made about what to do with the child. Since Albert and Mileva were not married, adoption seemed to be a solution, but they found this option difficult to accept. What happened to Lieserl remains open to speculation. Some suggest that she was indeed adopted, others that she died of scarlet fever just after birth. Still others suggest that she was put in a home for handicapped children because of the effects of the scarlet fever. Even today, Lieserl's fate remains a mystery. Andrea Gabor suggests that Albert never saw his daughter, since there is no record of his visiting Serbia while his wife was there. The decision on their daughter's future seems to have been left to Mileva, and it is obvious that it was an extremely difficult one for her to make. One possible reason for Albert's apparently casual attitude towards his daughter may be inferred from one of his letters to Mileva, dated December 12, 1901, in which he seems to reveal a gender preference by writing: "Be happy about my dear Lieserl, who I secretly (so Dolly [Mileva] doesn't notice) prefer to imagine a Hanserl" (Renn and Schulmann 1992, p. 66). Perhaps Albert would have preferred a son rather than a daughter.

With her pregnancy several months advanced, and without her beloved "Johnnie" (as she affectionately nicknamed Albert), who was away job hunting, Mileva began to study feverishly for the final examination, which she was to retake in 1901. She not only failed again but also dropped her dissertation. Her relationship with Weber had, it seems, deteriorated to the point of mutual indifference. The stigma of carrying an illegitimate child, in addition to Albert's being away looking for a job and the difficulties of the pregnancy, certainly were not conducive to tackling an examination with a clear mind. There were no counsellors to advise her and no friends to comfort her. If there were people she could confide in, her introverted personality likely would not have allowed her to express her feelings. The only recourse she felt was open to her was to transfer all her hopes and dreams onto Albert.

Mileva Marić had been only the fifth woman to be accepted into the prestigious male-dominated Eidgenössische Technische Hochschule, which was a remarkable achievement. When she lost respect for and alienated her thesis adviser, Weber, this was tantamount to committing academic suicide. How could this talented, bright young woman, with an abundance of potential, suddenly arrive at such a dismal state? Unlike Albert, she had not been a slow starter. Instead, like a true thoroughbred, she had come charging out of the block. She had overcome, with distinction, all the academic prejudice that women could encounter. Despite her physical affliction, she remained positive, and, in contrast to Albert's penchant for short-lived infatuations, she was capable of a sincere and durable love. Despite all her academic achievements, the ultimate recognition of receiving a degree was denied her, due in part to the mutually negative relationship with Professor Weber and Albert's exploitation of her time.

Mileva and Albert married in January 1902. She may have thought that she had chosen well, but did she choose wisely? Albert, for his part, was certain that he had chosen both wisely and well, and wrote to Mileva's father (as quoted in Troemel-Ploetz 1990, p. 418):

> I didn't marry your daughter because of the money, but because I love her, because I need her, because we are both one. Everything I have done and accomplished I owe to Mileva, she is my genial source of inspiration, my protective angel against sins in life and even more so in science. Without her I would not have started my work, let alone finish it.

This expression of love for, and reliance on, Mileva was not recognized by Albert later, when he had become universally famous for his work on the theory of relativity. There is little evidence that Mileva was ever named as co-author on any of the papers they worked on together.

Mileva gave birth to two sons, Hans Albert, in 1904, and Eduard, in 1910. Eduard was an extremely bright but sickly child who was prone to mood swings that eventually developed into violent fits. He was later diagnosed as manic-depressive, bordering on schizophrenic. Unlike his parents, he disliked mathematics and physics, and directed his studies towards literature and the arts. He developed a passion for poetry and music, and became quite a respected poet and an accomplished pianist. Both he and his brother had difficult relationships with their father and were never able to achieve a natural affinity with him. Hans Albert mentions of his father, "He needed to be loved himself. But almost the instant you felt the contact, he would push you away. He would not let himself go. He would turn off his emotions like a tap" (quoted in Highfield and Carter 1993, p. 223). When Hans Albert chose engineering as a profession, his father was bitterly disappointed. He would have preferred that his son study the pure sciences rather than adopt the practical, hands-on approach of engineering, and mentioned to Mileva in 1918: "What he is interested in isn't really important, even if it is, alas, engineering. One cannot expect one's children to inherit a mind" (quoted in Highfield and Carter 1993, p. 225). Eduard, meanwhile, spent his adult life in medical institutions and asylums and eventually succumbed to his numerous medical problems by suffering a fatal heart attack in 1964. Hans Albert survived his younger brother by eight years.

During his marriage, Albert's earlier tendency to violent outbursts reappeared in the form of physical abuse inflicted on his family. On at least one occasion, Professor Adolf Hurwitz's daughter Lisbeth, who had befriended the Einsteins, witnessed the result of a physical attack on Mileva: "Einstein had excused himself from the usual music evening with a vague reference to 'family matters.' The following day, Lisbeth and her mother visited Mileva and found her face badly swollen" (Highfield and Carter 1993, p. 153). By 1914, a separation, followed by divorce, seemed inevitable. Albert's philandering, his occasional violence, and his long absences from their home in Berlin sent Mileva into a deep depression. She decided to leave Berlin, a city she hated and where she had no friends, and returned to Zurich with the two boys. Albert periodically sent her money, but it was insufficient to cover the high cost of Eduard's medical treatment, and Mileva began to sink deeper into debt. She borrowed money from friends, and worked part time teaching piano and mathematics. According to Hans Albert, the separation was "very hard" on his mother, and he suggested that one possible reason for her melancholy was that she realized that her scientific life was over (Highfield and Carter 1993, p. 167). She had sacrificed her own career and ambitions and was totally dependent on Albert, even though he was becoming more and more remote from her.

In this respect, Milan Popović offers some insight into Einstein's personality that Popović himself predicts "will be troubling for those who only see Einstein as a saint," quoting a letter, dated September 8, 1916, from Einstein to Ellen Kaufler, who was Popović's grandmother and a close friend of Mileva's (in Popović 2003, p. 21):

> Separation from Mitsa [Mileva] was for me a question of life. Our life in common had become impossible, even depressing, but I could not say why. So I am giving up my boys whom I love so tenderly. During the two years of our separation, I have seen them twice; last spring I took a little trip with [Hans] Albert. To my great sorrow, I have found that my children do not understand my actions, that they feel a mute anger against me, and I find, although it hurts me, that it is better for them if their father does not see them any more. I shall be satisfied if they become useful and respected men. . . . I have great trust in the influence of their mother. . . . Mitsa's illness has made me sad, but fortunately she is now in full convalescence. Despite this interest on my part, she is and will remain always for me a severed limb. I shall never again approach her; I shall finish my days far away from her, feeling that it is absolutely necessary.

Popović's own reaction to this letter is one of disbelief. He describes as shocking, and his behaviour as a classic example of guilt brought on by his affair with Elsa. His verdict is that "Einstein narcissistically portrays himself as the victim. . . . He had abandoned his humanity at the cost of his family's happiness and his wife's well-being" (Popović 2003, p. 22).

Albert's parental distance and casual concern, not only for his now forgotten daughter, but later for his two sons, may surprise some readers, but even one of his most devoted admirers and biographers, Abraham Pais, recognizes Albert's noncommittal attitude when he writes (1982, p. 14) that Einstein "had no need to push the every day world away from him. He just stepped out of it whenever he wished."

The Einsteins' divorce was finalized on St. Valentine's Day, February 14, 1919, and a few months later Albert married his cousin Elsa Lowenthal. When he received the Nobel Prize for Physics, in 1922, he agreed as part of the divorce settlement that the prize money be given to his wife in full. It has been suggested that the reasons for Albert's generous payment were to ensure that Mileva would not cause him trouble by asking him for money, and to clear his conscience over his leaving Mileva for another woman and turning his back on his sons. With the prize money Mileva purchased three properties in Zurich, renting out two of them and making the third her residence. Gradually, however, Eduard's medical expenses became too much of a financial burden for her, and she was forced to sell both the rental properties. Her ownership of her home was eventually contested by Albert, but fortunately she retained power of attorney over the house.

Time was running out for Mileva. Her health was deteriorating rapidly, due in part to Eduard's constantly violent and destructive behaviour, and Albert's indifference to her circumstances. She had suffered several strokes, the final one leaving the left side of her body paralyzed. She died quietly and alone on August 4, 1948, at the age of seventy-three. There is no longer any trace of her gravesite in the Nordheim Cemetery in Zurich. Like Mileva Marić Einstein herself, it has simply disappeared.

Owing to Albert's mediocre beginnings, perhaps he could be categorized as an example of the classic "slow starter" who gradually achieves greatness. But was his work on the theory of relativity a singular achievement? Or did Mileva Marić contribute, if not in a major role, then at least as a collaborator, to Einstein's work? If she did, why has she been relegated to obscurity?

A particularly contentious debate arose in the world of physics after the publication of a biography of Mileva by a Serbian mathematician and physicist, Desanka Trbuhovic-Gjuric (1897–1983). This appeared in the former Yugoslavia in 1969, but the controversy did not begin until it became known in other countries through translations, in the early 1980s. Troemel-Ploetz suggests (1990, p. 417) that Trbuhovic-Gjuric's main purpose in writing the book was "to rescue Mileva Einstein Marić from oblivion, and write her into Serbian and scientific history. She knew that no man would do the job for Mileva Einstein Marić, whose own husband failed to give her the public recognition she deserved." It was the first biography to mention that Albert and Mileva had had an illegitimate child, which is now universally accepted as fact, but it also raised questions of credibility over some of its other assertions, since it lacked documented and verified sources. Trbuhovic-Gjuric herself claimed that she had recounted "small events which I learned from word of mouth, letters, and family members she would leave it to the reader to draw his or her own conclusion". (in Barnett 1998, p. 139)

It has been suggested that behind many a remarkable man there is an equally remarkable woman. Mileva Marić is, arguably, a prime example of such a woman. It can be suggested that Mileva initially was Einstein's equal. History has shown that the playing field has rarely been level for women, and Mileva seems to have been no exception to that rule. She was neither the first nor, sadly, the last woman to lose her enthusiasm for her academic interests and settle down into a subservient role in marriage.

When Albert Einstein was recognized for his work on the theory of relativity, he was seen to be endowed with the gift of genius. There was another talent he possessed that may have been overlooked: his ability to sense that spark of genius in others. His admiration for Marić's mathematical skills in their early years together at the Eidgenössische Technische Hochschule was made evident when he later admitted to

Mileva's brother Milos, and to several Serbian intellectuals: "I need my wife. She solves all the mathematical problems for me" (Barnett 1998, p. 164). Mileva may have solved many mathematical problems for Albert, but if so, her work was never recognized formally in any of his papers.

For several years before and during the birth of their second son Eduard, in 1910, the couple's marriage rested on tenuous grounds due to their constant separations, and Albert's reliance on Mileva's mathematical skills and research assistance began to diminish. He now turned to others for help with his mathematical problems. As Clark (1971, p. 155) points out, when he was working on the general theory of relativity, "he was aided by his old friend Marcel Grossmann, the former colleague of a dozen years before whose notes during student days had enabled him to skip mathematics and concentrate on physics. . . . It was on Grossmann that Einstein now leaned heavily for the mathematical support." For all the assistance and ideas Grossman supplied, he too was never included in Einstein's papers as a co-author, receiving only a small acknowledgement at the end of one paper.

Fifty-four love letters between Albert and Mileva were discovered in 1986 and published in 1992. They span a period of six years, from 1897, when the couple first met, to 1903, when Mileva was pregnant with Hans Albert. Several of the letters indicate that Mileva was indeed a collaborator on Einstein's research on the theory of relativity. Letter 25, perhaps the most telling of the series, has caused the greatest controversy in the scientific community. Written by Albert to Mileva, and dated March 27, 1901, it includes the sentence: "I will be so happy and proud when we can bring *our work on relative motion* to a successful conclusion" (in Renn and Schulmann 1992, p. 39, emphasis added). While some take the phrasing "our work on relative motion" as an obvious indication that Mileva worked with Albert on the theory, the editors of the letters take it to be a reference to Albert's discussions of the theory with his friends Michele Besso and Marcel Grossmann, while others interpret it as merely a throwaway use of the royal "we." However, the next sentence reads: "When I see other people [Grossmann and Besso] I can really appreciate how special you are" (in Renn and Schulmann 1992, p. 39) Did Einstein mean that, although he discussed the theory with others, Mileva's input really was in some sense different, even "special"?

Among the fifty-four letters, eleven written by Albert appear to make references to joint work: that is, Albert sometimes makes statements such as "I am working . . ." or "I have . . ." when referring to his research, but elsewhere he writes as if more than one person were involved. Consider these passages from other letters that seem to refer to Mileva's collaboration (with emphases added in each case). From Letter 20, dated August 8 or September 6, 1900: "On the investigation of the Thompson effect, I have again resorted to a different method which is *similar to your method*" (in Renn and Schulmann 1992, p. 30). From Letter 21, dated September 13, 1900: "My work seems pointless and unnecessary if not for the thought that you are happy with what I am and what I do. . . . I am also looking forward to working on *our* new papers. You must continue with your investigations" (in Renn and Schulmann 1992, pp. 31–32). From Letter 26, dated April 4, 1901: "He [Michele Besso] is interested in *our* research . . . he went on my behalf to see his uncle, Professor Jung . . . to give him *our* paper" (in Renn and Schulmann 1992, p. 41). Many readers have concluded that these letters indicate collaboration, production of papers, and joint research.

However, the physicists Abraham Pais and John Stachel agree with the editors of the letters that there was little, if any, collaboration between Albert and Mileva on the theory, and that the use of such phrases as "our work" or "our paper" "was meant chiefly to serve the emotional needs of the moment" (Holton 1994, p. 42) This implies that Albert felt that Mileva was in need of psychological or emotional support, and was including her in his work by way of consoling her. Is this conjecture on their part, or is one required to accept this as their conclusive answer? How do they know when a particular moment was emotional? Whose emotional state was being consoled, Albert's or Mileva's, or both?

In a similar vein, Roger Highfield and Paul Carter refer to Albert's letter to Mileva's father, cited above, in the following terms (1993, p. 110):

> These lines certainly have the same tone as much that the young Einstein wrote in his love letters (of which Trbuhovic-Gjuric had no knowledge). But we have already seen that these letters are suggestive less of the intellectual relationship between Einstein and Mileva than of their emotional one.

Unfortunately, the evidence that these letters are to be interpreted as expressions of emotion is as minimal, and as contested, as the evidence that they should be interpreted as indicating an intellectual relationship. It is also interesting to note Highfield and Carter's reference to Trbuhovic-Gjuric. Those who favour her interpretation of Mileva's life, for all its acknowledged inaccuracies, would argue that the love letters confirm her approach, suggesting that it should not have been dismissed *tout court*.

The controversy continues, and opinions on both sides require a certain amount of judgment and interpretation from biographers and other writers. Carol C. Barnett illuminates this difficulty when she writes (1998, p. 151): "Surprising and of more immediate importance is the fact that members or proponents of each critical camp are relying upon identical historical information, yet they arrive at differing opinions and interpretations." Why does the scientific community, in general, support her critics, most of whom are male? Perhaps the answer is a matter not only of who to believe but of what kind of people we choose to believe. If the authoritative voice of science is predominantly male, then the few female voices will find difficulty in being heard, or for that matter, valued or believed.

Perhaps, then, it is necessary to return once more to an established fact, rather than an unsupported opinion. It is an established fact that, historically, academia has been dominated by a male culture and has been an environment in which women have had to struggle for acceptance and to be treated as equals. When women such as Mileva dared to encroach on the preserve of science, they were often greeted with ridicule and hostility, as Andrea Gabor explains: "At a time when few women . . . dared break through the rigid barriers to pursue a higher education, [they] were often harassed by male students and faculty" (Gabor 1996, p. 5). As Troemel-Ploetz writes: "The general attitude was, and is, that women do not belong there, so there are no positive expectations for them in the heads of male professors" (1990, p. 422). Equal recognition for women in academia has often depended on the eye of the beholder. Since that eye was, and still is, predominantly male, and possibly obscured by bias and prejudice, then equity was, and still is, a nebulous construct.

The concept of equality for society in general is certainly an ideal to strive for, but, as with several other practical ideals, it tends to provoke endless philosophical debate, or end in utopia. Perhaps the first step for women towards acquiring equality in academia is to require that men and women, at all levels of learning and instruction, place appropriate value on a woman's intellectual ability. "Different but equal" might be replaced with "different but valued." Only when a woman's contribution to society, whether in academia or in domesticity, is recognized and valued can a certain level of equality be attained.

Regarding Mileva Marić, the struggle towards truth still continues. Perhaps in the future we shall possess the whole truth, and then Mileva, like many other women who have fallen into obscurity, will finally receive the recognition she deserves.

ROSALIND FRANKLIN (1920–1958)

Rosalind Franklin's body of work developed through three different scientific investigations. For each achievement, she acquired an international reputation. Her initial research was on the physical and chemical properties of coal and charcoal, her second achievement was her important contribution that helped in the discovery of the structure of DNA, and her third project was her research on various plant diseases, particularly the tobacco mosaic virus. Each of these achievements is worthy of discussion, but the focus here is on her DNA research, due in part to the controversy it caused after her early demise.

The winners of the Nobel Prize for the discovery of DNA were Francis Crick, James Watson, and Maurice Wilkins. The important contribution that Rosalind Franklin made was the beautiful X-ray photograph, numbered 51, that gave Crick and Watson the major breakthrough they needed to complete and confirm their model of the structure of DNA. Watson later expressed this eureka moment by declaring, "The instant I saw the picture, my mouth fell open and my pulse began to rush. The pattern was unbelievable" (Watson 1968, p. 167). The controversy arose over the question of how they obtained photograph 51. Did they misappropriate it? Or was it stolen from Rosalind's laboratory by a circuitous route with the assistance of Maurice Wilkins, a reluctant co-researcher on the project? Serious as this accusation is, this is not our only concern. Rather, it is how once again a woman's contribution to a major event is downplayed or ignored by the scientific community.

Rosalind Franklin was born on July 25, 1920, into a relatively wealthy family. Both her parents, Ellis and Muriel, were descended from well-established intellectual and professional Jewish lineages. Besides Rosalind, they had four other children, three boys and a girl. Rosalind and her siblings were raised by two women: their mother and Ada Griffiths, who was employed as their nanny. The nanny was never considered as an employee or a servant, but was in essence the children's confidante, nurse, referee, and emotional adviser, and was at the centre of their universe. She took care of their physical and day-to-day needs, and their mother took control of their material wants and their intellectual development.

An unusual but positive aspect of the children's upbringing was the parents' reluctance to stereotype their children. Whenever any object was purchased for learning purposes, whether in construction or in handicrafts, all the children were involved. This early hands-on experience later became beneficial to Rosalind in her professional life, when she had to perform delicate experiments in her laboratory.

Rosalind's well-defined genetic inheritance became obvious from an early age. She was described by one of her aunts as "alarmingly clever." She loved problem-solving and, according to her aunt, spent most of her free time solving arithmetic problems. It eventually became obvious to her parents that a career in one of the sciences was inevitable. Anne Sayre, a friend and biographer of Rosalind's, wrote that "Rosalind had in a very high degree the talents, the mental capacities, the kind of intelligence, which lend themselves happily to science . . . she was, and indisputably so, a dedicated scientist" (Sayre 1975, p. 26). However, her father—whose patriarchal position was never questioned, as was then considered the norm in most Jewish families—had misgivings about his daughter's desire to be a scientist. He felt that a scientific career was unsuitable for a woman and foresaw the difficulties she would have to confront in such a male-dominated profession.

From the age of six, Rosalind and her brother David attended a private school that offered a comprehensive grounding in history, literature, and arithmetic. In her eleventh year, she entered St Paul's Girls' School in Hammersmith (West London), a leading independent school founded in 1904, and then perhaps best-known for having the composer Gustav Holst on its staff. She was now a "Paulina" (also the title of the school magazine). Miss Ethel Strudwick, the High Mistress of St Paul's, insisted that all her pupils concentrate on personal development, have a strong work ethic, and live a useful life. Marriage, she asserted, should not be their main goal, and women should not be relegated to the home.

St Paul's was the ideal school for Rosalind's developing talents, because it stressed competitive sports along with academic excellence, in the English tradition of *Mens sana in corpore sano* ("A sound mind in a sound body"). She studied chemistry, physics, and mathematics diligently, and her outstanding performance earned her a school-leaving grant of thirty pounds a year for three years, which was quite a considerable sum for that time.

University now beckoned and Rosalind had the choice of attending either Girton College or Newnham College, then the two women's colleges of the University of Cambridge. She decided on Newnham, although she was well aware that the university refused to award degrees to women (until 1948), partly on the grounds that it would be pointless to do so, since women would marry and have children rather than use a degree to enter a profession. Rosalind's approach to this prohibition, and the stereotypes it was based on, was to ignore them: "Quite clearly, and consciously, she chose the career, not the marriage" (Sayre 1975, p. 53) This logical response expressed Rosalind's independent mindedness. Her research assistant, Raymond Gosling, later offered the following insight: "She didn't suffer fools gladly at all. You had to be on the ball, or you were lost in any discussion about anything" (quoted in Sayre 1975, p. 105).

Rosalind plunged into serious laboratory work, but she also attended stimulating lectures by leading scientists. One in particular was given by the Nobel laureate father and son W. H. Bragg and W. L. Bragg on X-ray crystallography. Rosalind became intrigued by this form of scientific research, which was to lead her on the path towards the discovery of the structure of DNA.

Perhaps it is necessary here to digress a little and offer some insight into Rosalind the woman. The reason for this will become clear later when the professional Rosalind is described. Much has been written about Rosalind's knowledge, or lack of knowledge, of "the facts of life." Even in her early teens, she admitted to a cousin that she had never been kissed and was not sure how babies were conceived. It is not clear whether her sexual naivety was caused by nature or nurture, although the evidence we have lies heavily on the side of nurture. During her early years, her family rarely expressed overt physical affection. She avoided taking biology at St Paul's and when, later in her career, she was asked by King's College, Cambridge, to apply biological methods in her X-ray work, she responded: "I am of course, most ignorant about all things biological—I imagine most X-ray people start that way" (quoted in Maddox 2002, p. 108). There is no surviving evidence that she had lovers, or even any form of sexual expression or contact. She was an attractive woman, described by a colleague as quite stunning when dressed up for a social event. There is no indication whatever that she was physically attracted to her own sex, though she did prefer the company of women rather than men, and her few close male friends were mostly Jewish and married. It was suspected, but never confirmed, that the Jewish Italian crystallographer, Vittorio Luzzati, was romantically involved with her, but another Jewish physicist, Simon Altman, a long-time friend of Luzzati's, stated with certainty that they were not lovers (Maddox 2002, p. 138). All this seems odd, because in many other aspects she was a very curious, adventurous, and passionate woman. She enjoyed travelling, hiking in the mountains, the theatre, the arts, novels, and French cuisine. She spoke French fluently, could write and converse adequately in Italian, and seems to have preferred Mediterranean culture, with its fervent political debate and animated expression, over her own staid and reserved English culture.

One unusual event occurred in North Wales. According to Rosalind's most recent biographer, Brenda Maddox, Rosalind and a female friend, holidaying there, met two Welsh foresters. The men gave them a lift to a lake and as it was a very hot day, they decided to swim. Not having bathing suits, they all stripped off and swam naked. This was repeated the second day, but on this occasion her friend disappeared with one of the men. On their return to the campsite, they found Rosalind waiting with the other man, annoyed and puzzled: she did not understand why they had wanted to be alone together (Maddox 2002, p. 80) For Rosalind, this activity *au naturel* elicited no sense of shyness or shame: it was hot, the lake was inviting, it made sense.

Rosalind was a complicated woman. She could be impulsive and adventurous on some occasions, but very naïve and closed minded at other times. Her impulsive side never flowed over into her professional life, but the dichotomy in her character later proved to be detrimental in the undeclared race towards the discovery of the structure of the DNA.

During her research on DNA, her social intercourse with various colleagues was at a minimal level, and with one colleague in particular, Maurice Wilkins, it became a battle of wills, verging at times on utter contempt for each other. She did not accept his ideas or those of others, nor was she too impressed with their methods. One had to be right and back it up with solid evidence. For Rosalind, there was no middle ground, no "thinking outside the box," no impulsive adventures. Instead, she preferred to work in isolation, thus making herself unaware of what was happening around her, and therefore an easy target for the unscrupulous and ambitious duo of Francis Crick and James Watson.

In 1942, Rosalind, then twenty-two years old, was appointed an assistant researcher at the British Coal Utilisation Research Association (BCURA). Her main task was to study the physical and chemical structure of coal. Her work was very timely as a contribution to the war effort, then in full force, and her tenure at the BCURA has been described as "quite remarkable" by Professor Peter Hersch of the University of Oxford: "She brought order into a field which had been previously in chaos" (quoted in Sayre 1975, p. 64). In only four years, she produced five outstanding papers, developed an international reputation, and submitted a thesis for which, in 1945, she was awarded a doctorate by Cambridge—women could still not receive bachelor's degrees, but doctorates were special cases.

After leaving the BCURA, in 1946, Rosalind left for France to hike in the Alps and reconnect with a close friend and confidant, Adrian Weill, who was a metallurgist. This trip turned out to be pivotal to her career. Weill introduced Rosalind to two noted crystallographers, Marcel Mathieu and Jacques Mering, whom she impressed with her delicate technique in laboratory work. Mering offered her a job in a French government laboratory of which he was director and suggested that, instead of her physical and chemical approach to the study of coal, he should teach her the X-ray diffraction method, so that she could study the materials internally. This was a dream come true for Rosalind. She loved France, and quickly came to value Mering, her newly found mentor, who was a Russian-born Jew, married, but with a mistress and a flair for attracting women. Rosalind became quite infatuated with him, and there is a suggestion that something physical may have occurred. Mering later recalled that "there had been 'something' between them, [and] that he loved her very much, but she took things 'seriously,' and was naïve and inexperienced" (Maddox 2002, p. 97).

With her solid reputation as an expert in X-ray crystallography, Rosalind was offered a job at King's College, London, which she accepted with apprehension, uncertain whether she could cope with the parochial English lifestyle after her experience of challenging and liberal debate in France. The post brought her into contact with J. T. Randall, the head of the college's Physics Department and its Director of Medical Research, who had a knack for obtaining grants and finding highly qualified people to work in his laboratory. He was also, apparently, "manipulative and not above playing people off against each other" (Maddox 2002, p. 133), and may have contributed, knowingly or not, to the difficulties between Franklin and Wilkins by assigning them both to work on DNA without clarifying who was to do what, or who was to be in charge. Initially, they worked separately and were cordial to each other, but a storm was about to break. Rosalind's dislike of parochial Englishmen was directed towards Wilkins, whom she saw as too stuffy, too weak, and too provincial. The sharing of ideas and research became a cat-and-mouse routine, and Rosalind's research assistant, Gosling, often had to act as a go-between.

Rosalind's two-sided personality has been described quite bluntly by one of her acquaintances, Norma Sutherland, who told Brenda Maddox that "her manner was brusque and at times confrontational . . . she aroused quite a lot of hostility among the people she talked to and seemed quite insensitive to this. But she was kindness itself to me" (Maddox 2002, p. 151). Maurice Wilkins was Rosalind's temperamental opposite: "Her speech came fast, his slow; he evaded one's gaze, . . . turning away so that the listener was left facing his back. She positively liked hot and heavy debate, he became expressionless and quiet" (Maddox 2002, p. 146). There is a suggestion that Wilkins too was half in love with her, and that his behaviour was a form of aggressive foreplay, but if it was, it failed miserably. Rosalind preferred strong and decisive men, like her father. Nevertheless, the research went on, slowly but steadily, as each contributed progressive clues to the daunting mystery confronting them.

Rosalind and Wilkins were in no particular hurry because they were under the impression that no one else was working on DNA. Sir Lawrence Bragg of the Cavendish Laboratory in Cambridge had told James Watson and Francis Crick to stop their research on DNA, and apparently there was a "gentlemen's agreement" between the Cavendish and King's College, London, that only King's would work on DNA (Sayre 1975, p. 115). However, Watson and Crick carried on their work regardless.

James Watson had been a remarkably bright student. He had entered university at the age of fifteen and received a doctorate in zoology from the University of Chicago at twenty-three. After reading Erwin Schrödinger's book *What is Life?* he had become fascinated with gene theory, especially with the problem of how genes replicate themselves from generation to generation: why didn't they simply die, like all other animate matter? This question also fascinated Rosalind Franklin. However, with her as with other women, Watson acted the stumbling, bumbling, awkward virgin he then was. His memoir *The Double Helix* (1968) makes it obvious (even if Watson may not have intended it to) that the complicated, sophisticated, and cultured Rosalind was far beyond his immature experience, and that his only recourse was to demean her, to the point where he had to be coerced into adding an epilogue in which he retracts his antagonistic description of her.

In contrast, when Watson and Crick met, they had clicked immediately. Watson's knowledge of biology and genetics, coupled with Crick's familiarity with X-ray crystallography, made the perfect fit for their research. Both men were ambitious and understood the tremendous importance of discovering the structure of DNA. Naively, unguardedly, or with ill intent, Wilkins showed Franklin's remarkable photograph 51 to Watson, without her knowledge. The picture immediately produced a eureka response, for he could see plainly that the structure of DNA was helical (hence the title of his later memoir). Watson and Crick set out in earnest to build a model to prove this hypothesis. Unlike Watson's and Crick's harmonious relationship, Franklin's and Wilkins's noncommittal behaviour impeded their progress, allowing Crick and Watson to surge ahead. Rosalind was not averse to model-building, but her highly disciplined approach to research made her cautious: she felt that she needed more evidence before a model could be developed. Thus her adventurous side was stifled, with disastrous results.

Wilkins later asserted, erroneously, that Rosalind Franklin was "antihelical," that is, she doubted whether the structure of DNA was a helix at all, let alone a double helix. Ann Sayre comments (1975, p. 128),

> If Wilkins was sceptical about the accuracy of Rosalind's experiments, or her measurements, the scepticism proved to be unjustified. What Wilkins himself has said is: "I looked at the photograph—that B-form picture—and there it was, you can see the helix right there on the picture—but she refused point blank to see it. She was definitely antihelical."

Yet Rosalind's own lecture notes from November 1951 confirm that she was not antihelical. After viewing photograph 51, she wrote, "Conclusion: Big helix in several chains, phosphates on outside, phosphate–phosphate into helical bonds disrupted by water. Phosphate links available to proteins" (quoted in Sayre 1975, p. 128).

Waiting in the background for all the necessary clues to come together, Watson and Crick knew that eventually Rosalind, with due diligence, would provide them, as Sayre describes (1975, p. 135): "In the spring of 1953, Crick and Watson produced their triumphant and correct model: the solution to the DNA problem. By that time, some essential clues had come in[to] their hands which had been provided by Rosalind's liking for hard facts."

To the winners go the spoils. Brenda Maddox observes that when the trio of Crick, Watson, and Wilkins received their Nobel Prizes, four years after Rosalind's death from cancer, "Rosalind's name *was* praised from the Nobel platform in Stockholm, but not by Watson, Crick, or Wilkins." Sayre sums up Rosalind's treatment quite forcefully and bluntly by citing a few lines from Robert Frost's poem "Kitty Hawk":

> Of all the crimes the worst
> Is to steal the glory, . . .
> Even more accursed
> Than to rob the grave.

Today's women researchers can learn a lesson from Rosalind's short life: do not isolate yourself, network with others, learn the unwritten rules, know who your allies *and* your foes are, and ensure that your good work is documented and is made visible.

———⁂———

EPILOGUE

Peter Frize, Nadine Faulkner and I hope to have demonstrated in this book that the predominant patriarchal system has, until recently, prevented the access of girls and women to higher education, to membership of academies, other learned societies, and professional associations, and to positions in the public domain. This has created a world of science without women, allowing an exclusively masculine perspective to develop and perpetuate itself.

However, in every historical era, there have been courageous women who overcame obstacles and were able to produce serious scientific work. Many were ridiculed by famous authors, and others were severely criticized for not focusing their attention and energy on their "womanly duties" as wives and mothers. Many of them have been ignored by historians, and/or their contributions have been attributed to men. It is true that women frequently took male pseudonyms in order to publish, or had male relatives sign their work, yet even when these men confirmed that the work was done by the women, some historians continue to perpetrate the original attribution and ignore the correction. A case in point is the contributions of Maria Winkelmann (see Chapter 4), whose husband first signed the report to the King and later insisted that his wife had discovered the comet while he was sleeping.

Throughout the ages, some women were determined to follow their interests, explore their talents, and make important contributions to the development of knowledge. These pioneers offer real inspiration for today's women, and men, demonstrating that, with support and encouragement, their goals could be set and achieved.

It has not been possible to analyze the materials presented in this book in depth, spanning as it does so many centuries and so many countries, principally, but not exclusively, the

United Kingdom, France, and Germany in Europe, as well as Canada and the United States. However, I hope that the materials presented will have stimulated the curiosity of readers, who can explore more deeply some of the important sources referenced here.

I have also shared some of my own experiences as an engineering student, when few women were in these classes, and as a chair for Women in Science and Engineering. The knowledge I acquired through these experiences made me want to look into the past in order to understand the present and to make suggestions for a better future, in which science and engineering would be more gender balanced at all levels, including of course decision-making ones. I have read many excellent books, and visited several useful websites, enabling me to cover a very broad historical period, albeit not in depth, but, I believe, sufficiently to excite people about this topic. I fervently hope that others will continue to search for solutions and share them through their own writings.

The most difficult issue discussed in this book is perhaps the "nature versus nurture" debate. Some people prefer to think that men and women can be the same if they are socialized in a similar way. Others prefer to think that the characteristics and attributes assigned to each gender are different, and should remain so, as long as both genders are valued and respected, and these attributes do not engender stereotypes, biases, and obstacles in the manner that they currently do. We have seen that some authors, such as Linda Jean Shepherd, believe that these differences will create better teams for solving problems of a scientific or technical nature. No matter which of these approaches the reader favours, the important point is to develop better understanding, esteem differences where they exist, and avoid misperceptions regarding the contributions and work dynamics of the various groups. This will make everyone feel part of the group, and better results will be achieved by diverse teams.

Progress has been slow. Part of the reason for this is the prevalence of beliefs in the myths discussed in this book. When we get rid of the stereotypes and false ideas about women's abilities and skills, and face squarely the fact that there still are biases and double standards in many fields, then it will be possible to create an atmosphere of respect and trust for women's true abilities that will lead to equality. The other major factor is the acceptance that the predominantly male view is not the only way to create new knowledge. The diversity of perspective that women can bring to research and development in engineering and science will undoubtedly benefit these fields. The world, to date, has mainly documented men's ideas, perspectives, and ways of doing things.

With the women's movement of the 1960s and 1970s, and the interest of some researchers, mainly in the field of women's studies, we are rediscovering women and their contributions in literature, mathematics, history, and science. We must profile these women and their work for generations to come, so that girls can think, "I can also do this," and boys can think, "Girls and women can also do this."

When more women joined the professions of law, and medicine, new streams of specialization developed, such as family law or family practice in medicine. It is therefore possible that, when science and engineering become more gender balanced, some aspects of the culture will change and new areas will emerge. What remains is to integrate women's values in more than just a few aspects of a field. It is vital that they permeate through all aspects of knowledge in order to have a real impact. Considering that feminine characteristics have not been considered in a positive light for many centuries, it will take some time for the majority of scientists and engineers, including many of the women, to understand their value and to see how they can add complementary and enriching approaches to such work. It is to be hoped that, as the message gets repeated often enough, some of the men and women in science and engineering will begin to think about what it means, and find ways to integrate these attributes into their own practice. When people involved in the science and engineering enterprise visibly value these feminine characteristics, then we can expect a change in the mainstream culture, and women who wish to be different will no longer fear the stigma that this currently carries. Then we can all live in a world designed by balanced teams of women and men—hopefully, a world of peace and prosperity.

Effecting change in attitudes and behaviour takes time: it cannot be done overnight. Equity is not just having equal numbers of women and men. It also means securing equal chances of success and career development. It means equality in the respect obtained from peers and employers. It means having a voice and important roles at meetings and conferences. It also means a world where harassment and violence have disappeared from women's lives.

Occupations in non-traditional fields generally offer good salaries. If more women feel comfortable in making these choices, they will achieve economic independence, which in turn will give them more control over their lives. Women must face challenges, discover their talents and skills, and believe in themselves.

They should also choose their life partners as carefully as their careers to ensure that the two are compatible. Women and men must be agents of change, each in their own way.

A FEW WORDS ON THE PROFESSION OF ENGINEERING

During my forty years as an engineer, there were times when I still asked myself why I stayed. This was particularly acute when I attended the annual meetings of professional engineering associations, since for many years they included events aimed at "the wives," and when I heard about the existence of an Engineers' Wives Association, and wondered how this could relate to the husbands of women engineers. However, the interesting work and the challenge of solving problems that would help humanity kept me active in the profession. My message to colleagues who are engineers is that the profession of engineering can benefit from integrating and valuing the attributes, approaches, and ideas that women bring to the field. Women can have excellent and rewarding careers in medicine, pharmacy, physical and occupational therapy, and many other fields. They will not flock to engineering unless they see a welcoming profession in which they can thrive and balance their personal lives with their work. If the leaders of the profession begin to see how diversity can benefit engineering, and if they have respect for differences, then women will finally take their rightful places in the ranks of the profession. The profession will be richer for it.

—⊱⊰—

APPENDICES

APPENDIX 1
ARGUMENTS ABOUT WOMEN'S NATURES

Table A1.1 Framework of forms of arguments about women's natures

Women's natures (minds, talents, abilities, or capacities)	Logic of position	Child bearing capacity	Structure of society	Philosophers' views that can be used to support position
A. The same	1. Men and women have the same talents/abilities/capacities. 2. Access to education is based on one's capacities. 3. Therefore women should have access to education.	Not relevant	Based on nature. Therefore women can occupy the same social positions as men.	Plato Averroes[1] Astell[1] Descartes[2] de Poullain Locke Mill[3]
B. Different (less than or inferior) usually attributed to the Ancient Greeks	4. Men and women have inherently different and unequal talents/abilities/capacities. 5. Access to education is based on one's capacities. 6. Therefore women should have a different education than men.	Not necessarily relevant	Based on nature. Therefore women cannot occupy the same social positions as men.	Aristotle Aquinas Tertullian James[4] Hume Kant Broca

Table A1.1 (*continued*)

Women's natures (minds, talents, abilities, or capacities)	Logic of position	Child bearing capacity	Structure of society	Philosophers' views that can be used to support position
Different (complementary) especially prevalent in the 17th to 19th centuries	7. Men and women have inherently different natural talents/abilities/capacities. 8. Access to education is based on one's capacities. 9. Therefore women should have a different education than men.	Relevant	Based on nature, so women cannot occupy the same social positions as men and vice versa.	Aristotle Tertullian Kant Hume Wollstonecraft[5]

Notes

[1] Averroes and Astell may not have envisioned a completely transformed society.

[2] Descartes himself did not argue for women's education, but de Poullain uses Cartesian arguments.

[3] Mill seems rather to base the structure of society on the principle that everyone, regardless of gender, birth or race ought to be treated equally so that his/her nature may have the greatest chance of optimum development.

[4] William James supported education for women, but his views nevertheless support the view that women should not be given access to higher education.

[5] Mary Wollstonecraft argues both for a radical social reform and for maintaining the general structure of society in which women occupy distinct but complementary social positions.

Table A1.2 How data is accounted for by the different positions

Position	Data	Interpretation	Philosopher
Women's nature essentially different from men's	skilled or intellectual women	an exception or anomaly, "women with beards" (Kant), odd and undesirable	Aristotle Aquinas James
(this includes both the view attributed to the Greeks that women are lesser men, and the NCT view that they are altogether distinct, but complementary)	frivolous, intuitive, emotionally swayed, or ignorant women	evidence of women's true nature	Hume Kant Rousseau
	Society	plays no significant role in forming individuals	
Women's nature equal to men's	skilled or intellectual women	evidence of women's true nature	Plato Averroes Astell Wollstonecraft (?)
		a product of appropriate nurturing	
	frivolous, intuitive, emotionally swayed, or ignorant women	a product of inappropriate nurturing	Descartes (?) de Poullain Mill
	Society	plays a significant role in forming individuals	

APPENDIX 2
DEGREES AWARDED
AND STUDENTS ENROLLED
IN SCIENCE AND ENGINEERING

Table A2.1 Changing proportions of undergraduate degrees awarded to women in engineering in the United States and Canada, selected years, 1966–2005

	Proportions in United States (%)	Change from previous selected year in United States (absolute numbers)	Proportions in Canada (%)	Change from previous selected year in Canada (absolute numbers)
1966	0.4			
1970	0.8	130		
1975	2.1	151	3.6	
1980	10.1	608	7.9	181
1985	14.5	89	10.8	53
1990	15.4	– 11	14.0	37
1995	17.3	10	18.9	54
2000	20.5	12	20.3	28
2001	20.1	– 2.4	20.6	6.3
2002	20.9	6.4	19.9	1.97
2003	20.3	2.2	19.2	–.003
2004	20.5	2.2	18.2	–4.1
2005	NA		17.5	–4.7

Sources: National Science Foundation, US (2007) and CCPE, Canada (2007).

Table A2.2 Distributions of female undergraduate students across disciplines in universities in Quebec, 1999 and 2005 (%)

	1999	2005
Health sciences	11.6	16.0
Science and engineering	17.0	13.5
Humanities and social sciences	22.0	21.9
Education	22.4	20.5
Administration	12.6	13.8
Arts	4.7	4.8
Literature	5.0	3.8
Law	3.7	3.7
Multidisciplinary studies	1.0	2.0

Table A2.3 Distributions of female students across disciplines in the first year of undergraduate study in universities in Quebec, 1999 and 2005 (%)

	1999	2005
Health sciences	9.9	15.4
Sciences and engineering	17.6	12.2
Humanities and social sciences	23.3	23.5
Education	22.1	19.6
Administration	11.7	13.6
Arts	5.3	5.0
Literature	4.8	4.2
Law	4.3	4.3
Multidisciplinary studies	1.0	2.2

Table A2.4 Distributions of male undergraduate students across disciplines in universities in Quebec, 1999 and 2005 (%)

	1999	2005
Health sciences	5.1	5.8
Science and engineering	45.8	44.5
Humanities	15.1	14.3
Education	8.4	8.1
Administration	15.1	17.4
Arts	3.5	3.5
Literature	2.5	1.8
Law	3.4	3.0
Multidisciplinary studies	1.0	1.7

Table A2.5 Distributions of male students across disciplines in the first year of undergraduate study in universities in Quebec, 1999 and 2005 (%)

	1999	2005
Health sciences	4.6	5.8
Science and engineering	44.9	38.2
Humanities	15.6	16.9
Education	8.3	10.0
Administration	14.4	17.2
Arts	4.4	4.0
Literature	2.5	2.0
Law	4.3	4.0
Multidisciplinary studies	1.1	1.9

Table A2.6 Proportions of women enrolled in scientific disciplines in universities in the United States, selected years, 1966–2001 (%)

	Physical sciences	Mathematics and computer sciences	Biology and agricultural studies	Earth, atmosphere and ocean sciences
1966	14.0	33.2	25.0	9.4
1970	14.5	36.1	24.1	10.2
1975	18.8	37.0	29.2	17.0
1980	24.0	36.4	39.1	23.8
1985	29.7	39.5	45.1	24.6
1990	32.2	35.8	48.2	27.9
1995	35.5	35.1	49.7	34.0
2000	41.1	32.7	55.8	40.0
2001	41.7	31.8	57.3	40.9

Source: U.S. National Science Foundation.

APPENDIX 3
NOMINATIONS AND ELECTIONS OF FELLOWS
OF THE U.S. INSTITUTE OF ELECTRICAL AND
ELECTRONICS ENGINEERS (IEEE), 1999–2009

Table A3.1

	Total numbers elected	Numbers of women elected	Numbers of women nominated	Proportions of women nominated who were elected (%)	Women elected as proportions of total numbers elected (%)
1999	245	13	21	61.9	5.3
2000	248	2	6	33.3	0.8
2001	260	5	17	29.4	1.9
2002	259	13	28	46.4	5.0
2003	260	14	32	43.8	5.4
2004	260	6	36	16.7	2.3
2005	268	17	46	37.0	6.3
2006	271	7	44	15.9	2.6
2007	268	18	48	37.5	6.7
2008	295	27	47	57.4	9.2
2009	302	19	46	41.3	6.3

Source: IEEE.

BIBLIOGRAPHY

Abercromby, David. (2003 [originally 1685]). *A Discourse of Wit.* Whitefish, MT: Kessinger Publishing.

Albisetti, James E. (2000). "Unlearned Lessons from the New World?: English Views of American Coeducation and Women's Colleges, c. 1865–1910." *History of Education* 29:5.

Albisetti, James E. (2004). "The French *Lycées de jeunes filles* in International Perspective, 1878–1910." *Paedagogica Historica* 40:1–2.

Alic, Margaret. (1986). *Hypatia's Heritage: A History of Women in Science from Antiquity through the Nineteenth Century.* Boston, MA: Beacon Press.

Anderson, Inger-Johanne Tveten. (2002). "The Social Construction of Female Engineers: A Qualitative Case Study of Engineering Education." PhD thesis, University of Saskatchewan.

Arditti, Rita. (1980). "Feminism and Science," in *Science and Liberation*, ed. Rita Arditti, Pat Brennan, and Steve Cavrack. Boston MA: South End Press.

Aristotle. (1953). *Generation of Animals*, text with tr. by A. L. Peck. London: Heinemann.

Aristotle. (1993). *De Anima: Books II and III, with Passages from Book I*, tr. D. W. Hamlyn. Oxford: Clarendon Press.

Arnal, Claude. (2006). "Villepreux-Power, Jeanne." Online at http://www.astr.ua.edu/4000WS/VILLEPREUX.html [consulted July 3, 2009].

Association Française des Femmes Ingénieurs. (2006). *Les femmes ingénieurs en France.* Paris: Association Française des Femmes Ingénieurs.

Averroes. (1974). *Averroes on Plato's* Republic, tr. Ralph Lerner. Ithaca, NY: Cornell University Press.

Axelrod, Paul. (1997). *The Promise of Schooling: Education in Canada, 1800–1914.* Toronto: University of Toronto Press.

Barnett, Carol C. (1998). "A Comparative Analysis of Perspectives of Mileva Marić Einstein." MA dissertation, Florida State University.

Baumgartner-Papageorgiou, Alice. (1982). *My Daddy Might Have Loved Me: Student Perceptions of Differences Between Being Male and Being Female.* Denver: Institute for Equity in Education, University of Colorado.

Belenky, Mary, Blythe Clinchy, Nancy Goldberger, and Jill Tarule. (1986). *Women's Ways of Knowing: The Development of Self, Voice, and Mind.* New York: Basic Books.

Bell, Linda A. (1983). *Visions of Women.* Clifton, NJ: Humana Press.

Bennett, Daphne. (1990). *Emily Davies and the Liberation of Women, 1830–1921.* London: André Deutsch.

Bijvoet, Maya. (1989). "Editor of Montaigne: Marie de Gournay (1565–1645)," in *Women Writers of the Seventeenth Century*, ed. Katharina M. Wilson and Frank J. Warnke. University of Georgia Press.

Bluestone, Natalie Harris. (1987). *Women and the Ideal Society: Plato's Republic and Modern Myths of Gender.* Amherst MA: University of Massachusetts.

Bluestone, Natalie Harris. (1994). "Why Women Cannot Rule: Sexism in Plato Scholarship," in *Feminist Interpretations of Plato*, ed. Nancy Tuana. University Park, PA: Penn State Press.

Bridenthal, R., S. M. Mosher Stuard, and M. E. Wiesner. (1998). *Becoming Visible: Women in European History.* 3rd ed. Boston, MA: Houghton Mifflin.

Bucciarelli, Louis L., and Nancy Dworsky. (1980). *Sophie Germain: An Essay in the History of the Theory of Elasticity.* Dordrecht: D. Reidel, and New York: Springer.

Burgess-Jackson, Keith. (2002) "The Backlash Against Feminist Philosophy," in *Theorizing Backlash: Philosophical Reflections on the Resistance to Feminism*, ed. Anita M. Superson and Ann E. Cudd. Lanham, Boulder, CO, New York, and Oxford: Rowman & Littlefield.

Burstall, Sara. (1909). *Impressions of American Education in 1908.* London and New York: Longman Green.

Byers, Nina. (1999). "Overview on Women's Education in England and the United States, 1600–1900." Online at http://www.physics.ucla.edu/~cwp/hist/WL.html [consulted July 3, 2009].

Canadian Committee on Women in Engineering (CCWE). (1992, April). *Women in Engineering: More than Just Numbers: Report of the Canadian Committee on Women in Engineering.* Ottawa: CCWE. Online at http://www.carleton.ca/cwse-on/webmtjnen/repomtjn.html [consulted July 3, 2009].

Canadian Council of Professional Engineers (CCPE). (2004). *Canadian Engineers for Tomorrow: Trends in Engineering Enrolment and Degrees Awarded, 1998–2002.* Ottawa: CCPE.

Canadian Council of Professional Engineers (CCPE). (2006). *Canadian Engineers for Tomorrow: Trends in Engineering Enrolment and Degrees Awarded, 2001–2005.* Ottawa: CCPE. Online at: http://www.engineerscanada.ca/e/pu_enrolment.cfm [consulted July 3, 2009].

Caplan, Paula J. (1993). *Lifting a Ton of Feathers: A Woman's Guide to Surviving Academic Life.* Toronto: University of Toronto Press.

Caruana, Claudia M., and Cynthia F. Mascone. (1992, January). "Women Chemical Engineers Face Substantial Sexual Harassment: A Special Report." *Chemical Engineering Progress.*

Catalyst. (2007). *The Double-Bind Dilemma for Women in Leadership: Damned if You Do, Doomed if You Don't.* New York: Catalyst Inc.

Cavendish, Margaret. (2000 [originally 1655]). *The World's Olio,* in *Paper Bodies: A Margaret Cavendish Reader,* ed. Sara Mendelson and Sylvia Bowerbank. Peterborough, ON: Broadview Press.

Cervantes, Miguel de. (1981 [originally 1604–17]). *Don Quixote: The Ormsby Translation,* tr. John Ormsby, ed. Joseph R. Jones and Kenneth Douglas. New York: W.W. Norton.

Charles, Frederic. (1991). "France," in *Girls and Young Women in Education: A European Perspective,* ed. Maggie Wilson. Oxford: Pergamon.

Clark, Ronald. (1971). *Einstein: The Life and Times.* New York: World Publishing.

Comenius, Jan Amos. (1896 [originally 1633–38]). *The Great Didactic,* tr. M. W. Keatinge. London: Adam & Charles Black (reprinted Whitefish, MT: Kessinger Publishing).

Coulter, Rebecca. (1993). *Gender Socialization: New Ways, New World.* Victoria: Research and Evaluation Branch of the Ministry of Women's Equality, Province of British Columbia, on behalf of the Working Group of Status of Women Officials on Gender Equity in Education and Training (Governments of Newfoundland, New Brunswick, Ontario, Manitoba, British Columbia, and Canada).

Cressy, David. (1980). *Literacy and the Social Order: Reading and Writing in Tudor and Stuart England.* Cambridge: Cambridge University Press.

Crombie, Gail, et al. (1999). *Bridging the Gender Gap in Computer Science Education.* Ottawa: Nortel Networks.

Crombie, Gail, and Patrick I. Armstrong. (1999). "Effects of Classroom Gender Composition on Adolescents' Computer-Related Attitudes and Future Intentions." *Journal of Educational Computing Research* 20:4.

Daly, Kevin (2007, April). "Gender Inequality, Growth and Global Ageing." *Goldman Sachs Global Economics Papers* 154. New York, London, and Frankfurt: Goldman Sachs Economic Research. Online as http://www.ftd.de/wirtschaftswunder/resserver.php?blogId=10&resource=globalpaper154.pdf

Davis, J. K. (1978). *Democracy and Classical Greece.* Hassocks: Harvester Press, and Atlantic Highlands, NJ: Humanities Press.

Dickson, David. (1984). *The New Politics of Science.* Chicago: Pantheon.

Disse, Dorothy. (2009). "Anna Maria van Schurman/Schurmann/Schuurman (1607–1678)." Online at http://home.infionline.net/~ddisse/schurman.html [consulted July 3, 2009].

DiTomaso, Nancy, and George F. Farris. (1992). "Diversity in the High-Tech Workplace: Diversity and Performance in R&D." *IEEE Spectrum* [a publication of the Institute of Electrical and Electronics Engineers] 29:6.

Doyal, Lesley. (2001). "Sex, Gender, and Health: The Need for a New Approach." *British Medical Journal* 323: 7320.

Easlea, Brian. (1981). *Science and Sexual Oppression: Patriarchy's Confrontation with Women and Nature.* London: Weidenfeld & Nicolson.

El Akkad, Omar. (2005, September 21). "Where Jobs Are and Students Aren't." *Globe and Mail.*

Esterson, Allen. (2007, November 12). "Einstein's Wife: PBS Fails the Test of Integrity." *Butterflies and Wheels.* Online at http://www.butterfliesandwheels.com/articleprint.php?num=263 [consulted July 3, 2009].

European Technology Assessment Network (ETAN). (2000). *Science Policies in the European Union: Promoting Excellence through Mainstreaming Gender Equality: A Report from the ETAN Expert Working Group on Women and Science.* Brussels: Research Directorate General of the European Commission. Online at http://cordis.europa.eu/improving/women/documents.htm [consulted July 3, 2009].

Fara, Patricia. (2004). *Pandora's Breeches: Women, Science and Power in the Enlightenment.* London: Pimlico.

Feynman, Richard P. (1972). "The Development of the Space–Time View of Quantum Electrodynamics: Nobel Lecture, December 11, 1965," in *Nobel Lectures, Physics 1963–1970*, ed. The Nobel Foundation. Amsterdam: Elsevier. Online at http://nobelprize.org/nobel_prizes/physics/laureates/1965/feynman-lecture.html [consulted July 3, 2009].

Foschi, Martha, Larissa Lai, and Kirsten Sigerson. (2004). "Gender and Double Standards in the Assessment of Job Applicants." *Social Psychology Quarterly* 57:4.

Fraser, Antonia. (1984). *The Weaker Vessel: Woman's Lot in Seventeenth-Century England.* London: Weidenfeld & Nicolson, and New York: Alfred A. Knopf.

Frize, Monique. (1995). "Eradicating Sexual Harassment in Higher Education and Non-Traditional Workplaces: A Model." Paper presented at the 11th Annual Conference of the Canadian Association Against Sexual Harassment in Higher Education (CAASHHE), Saskatoon, November.

Frize, Monique. (1998). "Impact of a Gender-Balanced Summer Engineering and Science Programme on Future Course and Career Choices." *Proceedings of the 1998 Women in Engineering Conference: Creating a Global Engineering Community through Partnerships.* West Lafayette, IN: Women in Engineering Programs and Advocates Network (WEPAN). Online as http://dpubs.libraries.psu.edu/DPubS?service=Repository&version=1.0&verb=Disseminate&handle=psu.wepan/1180129598&view=body&content-type=pdf_1# [consulted July 3, 2009].

Frize, Monique, et al. (1998). "Pinocchio's Nose, the Long and the Short of It: A Special Day for Grade 10 Female Students at Nortel." Presentation at the 11th Canadian Conference on Engineering Education, Halifax, July 5–7.

Frize, Monique, and Ruby Heap. (2001). "The Professional Education of Women Engineers in Ontario and Quebec (1920–1999): Enrolment Patterns." *Journal of the Japan Society of Mechanical Engineers.*

Gabor, Andrea. (1996). *Einstein's Wife: Work and Marriage in the Lives of Five Great 20th-Century Women.* New York: Penguin Books.

Gates, Barbara T., and Ann B. Shteir. (1997). *Natural Eloquence: Women Reinscribe Science.* Madison: University of Wisconsin Press.

Goldberg, Cary, and Scott Allen. (2005, March 18). "Researcher Admits Fraud in Grant Data." *Boston Globe*. Online at http://www.boston.com/news/nation/articles/2005/03/18/ researcher_admits_fraud_in_grant_data/ [consulted July 3, 2009].

Goldberg, Philip. (1968). "Are Women Prejudiced against Women?" *Transaction* 5.

Gornick, Vivian. (1983). *Women in Science: Portraits from a World in Transition*. New York: Simon &. Schuster.

Gould, Stephen Jay. (1981). *The Mismeasure of Man*. New York: W. W. Norton.

Gournay, Marie de. (1988). *Fragments d'un discours féminin*, ed. Elyane Dezon-Jones. Paris: J. Corti.

Hacker, Sally. (1981). "The Culture of Engineering: Women, Workplace and Machine." *Women's Studies International Quarterly* 4:3.

Hall, Roberta M., and Bernice R. Sandler. (1982). *The Classroom Climate: A Chilly One for Women*. Washington, DC: Association of American Colleges.

Hancock, Mark S., Rhian Davies, and Joanna McGrenere. (2004). "Focus on Women in Computer Science." Paper presented at the Western Canadian Conference on Computing Education, Okanagan University College, Kelowna, British Columbia, May 6 and 7. Online at http://www.cs.ubc.ca/wccce/program04/ Papers/mark.html [consulted July 3, 2009].

Harding, Sandra. (1986). *The Science Question in Feminism*. Ithaca, NY, and London: Cornell University Press.

Haslett, Beth, Florence L. Geis, and Mae R. Carter. (1992). *The Organizational Woman: Power and Paradox*. Norwood, NJ: Ablex.

Heap, Ruby. (2003). "Writing Them into History: Canadian Women in Science and Engineering since the 1980s," in *Out of the Ivory Tower: Feminist Research for Social Change*, ed. Andrea Martinez and Meryn Stuart. Ottawa: University of Ottawa Press.

Heeter, Carrie, Rhonda Egidio, Punya Mishra, and Leigh Graves Wolf. (2005). "Do Girls Prefer Games Designed by Girls?" Paper presented at the Digital Research Association Conference (DIGRA), Vancouver. Online at http://www.allacademic.com// meta/p_mla_apa_research_citation/0/1/4/9/4/pages14948/ p14948-1.php [consulted JUly 3, 2009].

Highfield, Roger, and Paul Carter. (1993). *The Private Lives of Albert Einstein*. New York: St. Martin's Press.

Holbrook, J. Britt. (2005). "Assessing the Science–Society Relation: The case of the U.S. National Science Foundation's Second Merit Review Criterion." *Technology in Society* 27. Online at http://www.ndsciencehumanitiespolicy.org/Library/library.html [consulted July 3, 2009].

Holton, Gerald. (1994, September). "Of Love, Physics, and Other Passions: The Letters of Albert and Mileva, Part 2." *Physics Today.*

Hufton, Olwen. (1995). *The Prospect Before Her: A History of Women in Western Europe.* New York: Alfred A. Knopf.

Hyde, Janet S., et al. (2008). "Gender Similarities Characterize Math Performance." *Science* vol. 321.

Ingram, Sandra, and Anne Parker. (2002, March). "The Influence of Gender on Collaborative Projects in an Engineering Classroom." *IEEE Transactions on Professional Communication* [a publication of the Institute of Electrical and Electronics Engineers] 45:1. Online (but with restricted access) at http://ieeexplore.ieee.org/xpl/tocresult.jsp?isnumber=21277&isYear=2002 [consulted July 3, 2009].

Jex-Blake, Sophia (1867). *A Visit to Some American Schools and Colleges.* London: Macmillan.

Keller, Evelyn. Fox (1985). *Reflections on Gender and Science.* New Haven, CT: Yale University Press.

Keller, Evelyn Fox, and Helen Longino. (1996). *Feminism and Science.* New York: Oxford University Press.

Kennedy, Hubert. (1987). "Maria Gaetana Agnesi," in *Women of Mathematics: A Bio-Bibliographic Sourcebook*, ed. Louise S. Grinstein and Paul J. Campbell. Westport, CT: Greenwood Press.

Kitto, H. D. F. (1951). *The Greeks.* Harmondsworth: Penguin Books.

Koblitz, A. H. (1987). "Sofia Vasileva Kovaleskaia (1850–1891)," in *Women of Mathematics: A Bio-Bibliographic Sourcebook*, ed. Louise S. Grinstein and Paul J. Campbell, ed. Westport, CT: Greenwood Press.

Kronk, Gary W. (2008). "Caroline Lucretia Herschel." Online at http://cometography.com/biographies/herschelc.html [consulted July 3, 2009].

Lafortune, Louise, and Claudie Solar. (2003). *Femmes et maths, sciences et technos.* Sainte-Foy, Quebec City: Presses de l'Université du Québec.

Larkin, June. (1994). *Sexual Harassment: High School Girls Speak Out.* Toronto: Second Story Press.

Laucius, Joanne. (2009, April 13). "Bridging the 'Gender Gap.'" *Ottawa Citizen.* Online at http://www.ottawacitizen.com/News/ Bridging+gender/1490384/story.html.

Light, R. J. (1990). *The Harvard Assessment Seminars: Explorations with Students and Faculty about Teaching, Learning, and Student Life: First Report.* Cambridge MA: Harvard University Graduate School of Education.

Locke, John. (1980). *The Second Treatise of Government.* Indianapolis, IN: Hackett.

Lougee, Carolyn C. (1999). "'Its Frequent Visitor': Death at Boarding School in Early Modern Europe," in *Women's Education in Early Modern Europe: A History, 1500–1800,* ed. Barbara J. Whitehead. New York and London: Garland Publishing.

Lupart, Judy L., and M. Elizabeth Cannon. (2000). "Gender Differences in Junior High School Students towards Future Plans and Career Choices." Paper presented at the National Conference for the Advancement of Women in Engineering, Science and Technology, St. John's. Online as www.mun.ca/ cwse/Lupart,Judy.pdf [consulted July 3, 2009].

Macaulay, Catherine. (1994 [originally 1790]). *Letters on Education.* Oxford and New York: Woodstock Books.

Maddox, Brenda. (2002). *Rosalind Franklin: The Dark Lady of DNA.* London and New York: HarperCollins.

Maisel, Merry, and Laura Smart. (1997). "Sophie Germain, Born: Paris, April 1, 1776, Died: Paris, June 26, 1831: Revolutionary Mathematician," in *Women in Science: A Selection of 16 Significant Contributors.* San Diego, CA: San Diego Supercomputer Center. Online at http://www.sdsc.edu/ ScienceWomen/germain.html [consulted July 3, 2009].

Makin, Bathsua. (1673). *An Essay to Revive the Ancient Education of Gentlewomen, in Religion, Manners, Arts & Tongues, with an Answer to the Objections against This Way of Education.* London: Thomas Parkhurst. Online at http://www.pinn.net/ ~sunshine/book-sum/makin1.html [consulted July 3, 2009].

Margolis, Jane, and Allan Fisher. (2001). *Unlocking the Clubhouse: Women in Computing.* Cambridge, MA: MIT Press.

Martin, J. R. (2000). *Coming of Age in Academe.* New York and London: Routledge.

Massachusetts Institute of Technology (MIT) (1999, March). "A Study on the Status of Women Faculty in Science at MIT." *MIT Faculty Newsletter,* 11:4. Online at http://web.mit.edu/fnl/ women/women.html [consulted July 3, 2009].

McDill, Moyra. (2007). "Membership Profiles in the International Council of Academies of Engineering and Technological Sciences (CAETS)," in *WEPAN 2007 National Conference Proceedings: Imagining the Future of Engineering*, ed. Mary R. Anderson-Rowland. Toronto: X-CD Technologies. Online as http://www.x-cd.com/wepan07/WEPAN2007_0027.pdf.

McDill, Moyra, and Shirley Mills. (2002). "Youth Participation Trends in Engineering and Sciences," in *Proceedings of the 12th International Conference of Women Engineers and Scientists (ICWES)*, Ottawa, July.

McLellan, James E. (1985). *Science Reorganized: Scientific Societies in the Eighteenth Century.* New York: Columbia University Press.

Merchant, Carolyn. (1980). *The Death of Nature: Women, Ecology and the Scientific Revolution.* London: Wildwood.

Mill, John Stuart. (2001). *The Subjection of Women*, ed. Edward Alexander. Edison, NJ: Transaction Publishers.

Mill, John Stuart, Harriet Taylor Mill, and Helen Taylor. (1994). *Sexual Equality: A John Stuart Mill, Harriet Taylor Mill, and Helen Taylor Reader*, ed. Ann P. Robson and John M. Robson. Toronto: University of Toronto Press.

Moody, JoAnn. (2004). *Faculty Diversity: Problems and Solutions*. New York: Routledge.

Mozans, H. J. (1974 [originally 1913]). *Women in Science.* Cambridge, MA, and London: MIT Press.

Murray, Margaret A. M. (2001). *Women Becoming Mathematicians: Creating a Professional Identity in Post-World War II America.* Cambridge, MA: MIT Press.

Natural Science and Engineering Research Council (NSERC). (1996). *Report: Increasing Participation of Women in Science and Engineering Research.* Ottawa: NSERC.

Nies, Allison. (1999). "Laura Bassi: A Physicist Supported by the Church." Online at http://www.hypatiamaze.org/laura/bassi.html [consulted July 3, 2009].

Natural Sciences and Engineering Research Council. (June 2006). EKOS Research Associates Inc., Evaluation of the University Faculty Awards Program.

Nobel Foundation. (2008) "Women Nobel Laureates." Online at http://nobelprize.org/nobel_prizes/lists/women.html [consulted July 3, 2009].

Noble, David F. (1992). *A World without Women: The Christian Clerical Culture of Western Science.* New York: Alfred A. Knopf.

Noordam, Bart, and Patricia Gosling. (2007, July 20). "Mastering Your PhD: Relating to Your Co-Workers' Personality Types." *Science Career Magazine*. Online at http://sciencecareers. sciencemag.org/career_development/previous_issues/articles/ 2007_07_20/caredit_a0700104/(parent)/68 [consulted July 3, 2009].

O'Connor, John J., and Edmund F. Robertson. (1996, December). "Marie-Sophie Germain," in *The MacTutor History of Mathematics*. St Andrews, Fife: University of St Andrews. Online at http://www-gap.dcs.st-and.ac.uk/~history/Biographies/ Germain.html [consulted July 3, 2009].

O'Day, Rosemary. (1982). *Education and Society, 1500–1800*. London: Longman.

Ogilvie, Marilyn, and Joy Harvey, ed. (2000). *The Biographical Dictionary of Women in Science: Pioneering Lives from Ancient Times to the Mid-20th Century*. 2 vols. New York: Routledge.

Osler, Margaret J., ed. (2000). *Rethinking the Scientific Revolution*. Cambridge: Cambridge University Press.

Pais, Abraham. (1982). *Subtle is the Lord: The Science and Life of Albert Einstein*. New York: Oxford University Press.

Paludi, M. A., and L. A. Strayer. (1985). "What's in an Author's Name?: Differential Evaluations of Performances as a Function of Author's Name." *Sex Roles* 12:3–4.

Paracelsus. (1951). *Paracelsus: Selected Writings*, ed. Jolande Jacobi. New York: Pantheon.

Patterson, E. (1987). "Mary Fairfax Greig Somerville," in *Women of Mathematics: A Bio-Bibliographic Sourcebook*, ed. Louise S. Grinstein and Paul J. Campbell. Westport, CT: Greenwood Press.

Perry, Ruth. (1986). *The Celebrated Mary Astell: An Early English Feminist*. Chicago: University of Chicago Press.

Perry, Ruth. (1999). "Radical Doubt and the Liberation of Women," in *Feminist Interpretations of René Descartes*, ed. Susan Bordo. University Park, PA: Penn State Press.

Phillips, Patricia. (1990). *The Scientific Lady: A Social History of Women's Scientific Interests, 1520–1918*. London: Weidenfield & Nicolson.

Piaget, Jean. (1993). "Jan Amos Comenius (1592–1670)." *Prospects* 23:1/2. Online at http://www.ibe.unesco.org/en/services/ publications/thinkers-on-education.html [consulted July 3, 2009].

Plato. (1974). *The Republic*, tr. H. D. P. Lee. 2nd ed. Harmondsworth: Penguin Books.

Popovič, Milan. (2003). *In Albert's Shadow: The Life and Letters of Mileva Marič, Einstein's First Wife.* Baltimore, MD: Johns Hopkins University Press.

Prentice, Alison L., and Susan Houston, ed. (1975). *Family, School and Society in Nineteenth-Century Canada.* Toronto: Oxford University Press.

Rappaport, K. D. (1987). "Augusta Ada Lovelace (1815–1852), in *Women of Mathematics: A Bio-Bibliographic Sourcebook,* ed. Louise S. Grinstein and Paul J. Campbell. Westport, CT: Greenwood Press.

Ray, Sheri Graner. (2003). *Gender-Inclusive Game Design: Expanding the Market.* Rockland, MA: Charles River Media.

Renn, Jürgen, and Robert Schulmann, ed. (1992). *Albert Einstein, Mileva Marić: The Love Letters,* tr. Shawn Smith. Princeton, NJ: Princeton University Press.

Robinson, G. J. and J. S. McIlwee. (1991). "Men, Women, and the Culture of Engineering." *Sociological Quarterly* 32:3.

Rogers, Pat. (1988). "Gender Differences in Mathematical Ability—Perceptions versus Performance." Paper presented at the 6th International Conference on Mathematical Education (ICME 6), Budapest.

Rose, Hilary. (1994). *Love, Power, and Knowledge: Towards a Feminist Transformation of the Sciences.* Cambridge: Polity Press, and Bloomington and Indianapolis: Indiana University Press.

Rosser, Sue V. (1990). *Female-Friendly Science: Applying Women's Studies Methods and Theories to Attract Students.* New York: Teachers College Press, Columbia University.

Rosser, Sue V. (1997). *Re-Engineering Family-Friendly Science.* New York: Teachers College Press, Columbia University.

Rosser, Sue V. (2004). *The Science Glass Ceiling: Academic Women Scientists and the Struggle to Succeed.* New York: Routledge.

Rossiter, Margaret. (1982). *Women Scientists in America: Struggles and Strategies to 1940.* Baltimore, MD: Johns Hopkins University Press.

Rousseau, Jean-Jacques. (1911 [originally 1762]). *Émile, ou, de l'éducation,* tr. Barbara Foxley. New York: E. P. Dutton and London: J. M. Dent.

Ruivo, Beatriz. (1987). "The Intellectual Labour Market in Developed and Developing Countries: Women's Representation in Scientific Research." *International Journal of Science Education* 9:3.

Sadker, David, and Myra Sadker. (1994). *Failing at Fairness: How America's Schools Cheat Girls.* New York: Charles Scribner's Sons.

Sandler, Bernice R. and Robert J. Shoop, ed. (1997). *Sexual Harassment on Campus: A Guide for Administrators, Faculty, and Students.* Boston, MA: Allyn & Bacon.

Sayre, Ann. (1975). *Rosalind Franklin and DNA.* New York: W. W. Norton.

Schiebinger, Londa. (1989). *The Mind Has No Sex?: Women in the Origins of Modern Science.* Cambridge, MA: Harvard University Press.

Schiebinger, Londa. (1999). *Has Feminism Changed Science?* Cambridge, MA: Harvard University Press.

Schools Inquiry Commission (1867–68), *Report*, vol. 1, chapter 6. Online at http://www.dg.dial.pipex.com/documents/docs1/sicr-girls.shtml [consulted July 3, 2009].

Schwartz, Felice N., and Jean Zimmerman. (1992). *Breaking With Tradition: Women and Work, the New Facts of Life.* New York: Warner Books.

Sévigny, J., and C. Deschênes. (2007). *Évolution des effectifs étudiants universitaires au Québec, 1999 à 2005—Analyse des données du MÉLS.* Quebec City: Chaire CRSNG/Alcan pour les femmes en sciences et génie au Québec and AFFESTIM.

Shepherd, Linda Jean. (1993). *Lifting the Veil: The Feminine Face of Science.* Boston, MA: Shambhala.

Shiva, Vandana. (1989). *Staying Alive: Women, Ecology and Development.* London: Zed Books.

Singer, Peter. (1999 [originally 1974]). "All Animals Are Equal," in *Bioethics: An Anthology*, ed. Helga Kuhse and Peter Singer. Oxford: Blackwell Publishers.

Singh, Simon. (1997). *Fermat's Enigma: The Epic Quest to Solve the World's Greatest Mathematical Problem.* New York: Walker and Company.

Smith, Janet Farrell. (1983). "Plato, Irony, and Equality." *Women's Studies International Forum* 6:6; reprinted in *Feminist Interpretations of Plato*, ed. Nancy Tuana. University Park, PA: Penn State Press.

Sonnert, G., and G. Holton. (1996, January–February). "The Career Patterns of Men and Women Scientists." *American Scientist*.

Sorensen, K. H. (1992). "Towards a Feminized Technology?: Gendered Values in the Construction of Technology." *Social Studies of Science* 22:1.

Sorensen, K. H., and A. J. Berg. (1987). "Genderization of Technology among Norwegian Engineering Students." *Acta Sociologica* 30:2.

Statistics Canada. (2005, October). *The Daily: University Enrolment.* Ottawa: Statistics Canada.

Stephans, Nancy Leys. (1996). "Race and Gender," in *Women in Science,* ed. Evelyn Fox Keller and Helen E. Longino. Oxford and New York: Oxford University Press.

Stock, Phyllis. (1978). *Better than Rubies: A History of Women's Education.* New York: Putnam.

Sutton, Geoffrey V. (1995). *Science for a Polite Society: Gender, Culture, and the Demonstration of Enlightenment.* Boulder, CO: Westview Press.

Teague, Frances. (1988). *Bathsua Makin, Woman of Learning.* Lewisburg, PA: Bucknell University Press, and London: Associated University Presses.

Tee, G. J. (1987). "Gabrielle-Émilie le Tonnelier de Breteuil, Marquise du Châtelet (1706–1749)," in *Women of Mathematics: A Bio-Bibliographic Sourcebook,* ed. Louise S. Grinstein and Paul J. Campbell. Westport, CT: Greenwood Press.

Tobias, Sheila. (1990). *They're Not Dumb, They're Different: Stalking the Second Tier.* Tucson, AZ: Research Corporation.

Tonso, Karen L. (1997, Spring). "Violence(s) and Silence(s) in Engineering Classrooms." *Advancing Women in Leadership Journal* 1:1.

Tonso, Karen L. (2006, January). "Teams That Work: Campus Culture, Engineer Identity, and Social Interactions." *Journal of Engineering Education.*

Toole, Betty. (2006). "Ada Byron, Lady Lovelace (1815–1852)." Online at http://www.cs.yale.edu/homes/tap/Files/ada-bio.html [consulted July 3, 2009].

Trbuhovic-Gjuric, Desanka. (1991 [originally 1969]). *Mileva Einstein: Une vie,* tr. of *U senci Alberta Ajnstajna.* Paris: Éditions des Femmes.

Troemel-Ploetz, Senta. (1990). "Mileva Einstein-Marić, the Woman Who Did Einstein's Mathematics." *Women's Studies International Forum* 13:5.

Tuana, Nancy, ed. (1994). *Feminist Interpretations of Plato.* University Park, PA: Penn State Press.

U. S. National Science Foundation. (2004). "Science and Engineering Degrees, 1966–2004." Online at http://www.nsf.gov/statistics/ nsf07307/ [consulted July 3, 2009].

U. S. National Science Foundation. (2007). "Science and Engineering Statistics." Online at http://www.nsf.gov/statistics/ [consulted July 3, 2009].

Valian, Virginia. (1998). *Why So Slow?: The Advancement of Women.* Cambridge, MA: MIT Press.

Valian, Virginia (2007). *Tutorials for Change: Gender Schemas and Science Careers.* New York: Hunter College of the City University of New York. Online at http://www.hunter.cuny.edu/genderequity/equitymaterials.html.

Van Beers, Anne M. (1996). "Gender and Engineering: Alternative Styles of Engineering." MA thesis, University of British Columbia.

Vickers, M. H., H. L. Ching, and C. B. Dean. (1995). "Do Science Promotion Programs Make a Difference?" *Proceedings of the More than Just Numbers Conference* (Fredericton, New Brunswick, May 10–12).

Wajcman, Judy. (2004). *TechnoFeminism.* Cambridge: Polity Press.

Walker, Evan H. (1989, February). "Did Einstein Espouse His Spouse's Ideas?" *Physics Today.*

Waters, Kristin, ed. (2000). *Women and Men Political Theorists: Enlightened Conversations.* Malden, MA, and Oxford: Blackwell Publishers.

Watson, James D. (1968). *The Double Helix: A Personal Account of the Discovery of the Structure of DNA.* New York: Atheneum.

Wernersson, Inga. (1991). "Sweden," in *Girls and Young Women in Education: A European Perspective*, ed. Maggie Wilson. Oxford: Pergamon.

Wertheim, Margaret. (1995). *Pythagoras' Trousers: God, Physics, and the Gender Wars.* New York: Random House.

Westfall, R. S. (1971). *The Construction of Modern Science: Mechanisms and Mechanics.* New York: Wiley.

Wharton, Etta. (2001). *Where We Are and Where We Need to Go.* Toronto: Professional Engineers of Ontario. Online as http://www.ospe.on.ca/pdf/weac_ew_report_2001.pdf [consulted July 3, 2009].

Whitehead, Barbara J., ed. (1999). *Women's Education in Early Modern Europe: A History, 1500–1800.* New York: Garland Publishing.

Wilson, Maggie, ed. (1991). *Girls and Young Women in Education: A European Perspective.* Oxford: Pergamon.

Wollstonecraft, Mary. (1975 [originally 1792]). *A Vindication of the Rights of Woman, with Strictures on Political and Moral Subjects.* Harmondsworth: Penguin Books, 1988.

Wood, Shaunda L. (1999). "Family, Home, and the School Environment's Influence on Gifted Girls' Perceptions of Choice to Take Extracurricular Science Classes." MA thesis, University of Ottawa. Online at https://www.ruor.uottawa.ca/en/handle/10393/8759 [consulted July 3, 2009].

Young, Michael, ed. (1971). *Knowledge and Control.* London: Macmillan.

INDEX

Composed in Times New Roman PS 10 on 13

Printed and bound in Canada